PRACTICE OF DESALINATION

PRACTICE OF DESALINATION

Edited by Dr. Robert Bakish

NOYES DATA CORPORATION

Park Ridge, New Jersey   London, England

1973

628.167
P88

87973
Mar. 1974

FOREWORD

This book is based upon a series of papers presented
by various authorities at St. Croix, U.S. Virgin Islands,
during December 1971.

This is the first publication under the auspices of the
St. Croix Corrosion Installation West Indies Laboratory,
College of Science and Engineering, Fairleigh Dickinson
University.

While the course was given in December 1971, the book
as here constituted represents the present state of the
art of the United States desalination industry in both
its underlying principles and their applications. The
greater portion of this material should remain valid for
some time to come.

CONTENTS

INTRODUCTION

This volume is the record of a series of lectures, the first attempt by the St. Croix Corrosion Installation, West Indies Laboratory of the College of Science and Engineering, Fairleigh Dickinson University, to become involved, and hope fully to contribute to the solving of some of the problems of the Caribbean region.

Though perhaps the fact is neither widely known nor appreciated, the availability of potable water is tied intimately to the growth of the region. Inasmuch as ground water facts for the region do not offer much hope for improvement in the potable water situation, it is through desalination and through efficient and economical desalination alone that the potable water supply will increase and make possible the growth of the region. While the "fuel" for growth in the Caribbean is potable water, and while it must be provided if the region is to prosper, those involved must oversee that it is provided without endangering the region's ecology.

The objective of this presentation was to review the progress of desalination technology and to see which applications would be suitable for use in solving the desalination problems of the Caribbean. It is indeed regrettable that neither the wide range of questions asked during the course, nor the many answers provided could be made part of this record. The same is unfortunately true for the substance of the round table discussions. This information was particularly germane to the region.

The content of this volume presents the substance of the presentations given, but not quite in the order that they were in the series. It begins with a brief summary of desalination processes and considers the important aspects of operational desalination processes and some of the more promising approaches in the development stage in some detail. Presentations dealing with materials of

1

construction and their behavior in desalination installations follow, and those dealing with the economics of desalination operations at different output levels and personnel training complete the presentations.

This was the first involvement of the undersigned in the field of desalination and as such my comments could well be disputed by veterans in the field. To me, however, the desalination business in the Caribbean has suffered most substantially from the lack of contacts and opportunity for exchange of experiences, problems and solutions among operators. I feel also that lack of appreciation of the fact that a desalination plant is basically a chemical operation and to be treated with corresponding care and precision has not helped the problem.

Another basic problem preventing carefree operations has been the lack of rugged and continuous duty monitoring instrumentation. It is certain that effort on the part of those in instrumentation to develop such devices will bring handsome returns, not only to those in instrumentation but to the desalination field as a whole.

Last, but not least, the problem indicates a basic failure in providing adequate education and training of the operating personnel on the substance of the operations. True, while many training programs will be handicapped by lack of adequate background in its trainees, at this juncture of time we should be in a position to remedy this situation.

It is the hope that the St. Croix Installation will continue and improve. The courses to follow and the personal relations developed during these courses, I am sure will begin to bear fruit in better communications, problem solving and more efficient operations. We trust that we will continue to be favored with the cooperation of those in desalination and related industries, as ultimately we all stand to gain from progress in desalination.

The Caribbean region is one of many nations and various ethnic origins. It is my belief that through cooperation in desalination work aimed at providing one of the essential components for growth of the region's potable water, a wonderful opportunity exists for establishing sound relations which should lead to more intimate cooperation in other technological problems facing the region.

While the undersigned initiated the activities under the auspices of the St. Croix Corrosion Installation which led to this series of lectures, the series and this volume would never have become a reality were it not for the generous contributions of time and effort by the participants. Special thanks are due to the Office of Saline Water and to Mr. G.W. O'Meara, its Director, who generously provided information and suggestions during the organizational period, as well as three most competent and eloquent speakers for the course. Thanks are due to the Aluminum Association and the Copper Development Association, the metals producers and equipment manufacturers who presented talks and contributed sections of this volume. The undersigned is particularly indebted to

Mr. Richard Ahlgren of Aqua-Chem, Inc. and Mr. Bruce Watson of Grace Chemical Company for their willingness to prepare presentations on extremely short notice to replace speakers originally scheduled. Thanks are due to Mrs. Phyllis Isenberg for her efforts both during the organization of the course and in the preparation of the manuscript.

Last, but not least, thanks are due to the staff of the West Indies Laboratory, and in particular to Mr. Lowell Bingham, its Resident Manager, whose resourcefulness and efforts above and beyond the call of duty made it possible for him to provide all that was asked of him.

Dr. R. Bakish

DESALINATION PROCESSES — A BIRD'S EYE VIEW

Dr. R. Bakish
Fairleigh Dickinson University

ABSTRACT

Desalination involves a variety of principles and processes derived from these principles. The paper to follow takes a bird's eye view of all of these and presents highlights of both principles and processes.

INTRODUCTION

Seawater can be desalinated by a steadily growing number of approaches. Some of these have been known for centuries, others have been introduced in recent years, and yet others are to be developed. In the process of developing a viable desalination technology, progress has been no different than in the development of any modern technology. Under the steady pressure of need, different processes are evaluated and gradually pass from the laboratory through pilot plant stages to commercial size operations. As in the development of other technologies, here also expensive mistakes and economic disappointments go hand in hand with progress.

Process difficulties which delay laboratory progress are quite different from those which ultimately determine the profitability or, for that matter, even the survival of a commercial process in the stringently competitive economic climate prevailing today. Success of a process rests on its economic viability under prevailing conditions. To determine it, we must consider capital investments, cost of processing, yield rates, and costs of operations including maintenance of the installation. While these will present the real costs of an operation, they alone do not determine the viability of the operation as depending on circumstances; a $1.75/1,000 gallon cost might be entirely too expensive in one case, while on the other hand, a cost of $7.00/1,000 gallons though

4

appearing exorbitant, might be considered acceptable in another case. The whole matter of desalination economics will be considered later in this volume in considerable detail.

Let us now look at the four basic desalination approaches. The first and oldest of these, distillation, has given birth to five basic modifications which are competing for commercial success. These are the vertical tube, the multistage flash, the multieffect multistage, the vapor compression and the solar humidification.

The second approach utilizes membranes and has three processes competing for commercial success. They are electrodialysis, transport depletion and reverse osmosis.

The third approach based on crystallization has also three processes based on it: the vacuum freezing vapor compressor, the secondary refrigerant freeze, and the hydrate formation.

The fourth and last desalination approach, the chemical one, has only one process, i.e., the ion exchange process competing for a place in desalination technology.

DISTILLATION

This is undoubtedly the oldest principle known for the separation of fresh water from the seawater. The process simply requires the boiling of the seawater and condensing the vapor to give the product, potable water. To accomplish this one must provide thermal energy to generate the steam in order to be able subsequently to condense it. This boiling point is a function of pressure: at sea level it is 212°F. and becomes lower at pressures below one atmosphere.

If one is to have an efficient distillation process, one must use at least part of the heat recovered on condensation of the steam for the production of more steam. In addition, one must keep heat transfer surfaces free from scale. The presence of calcium sulfate, whose solubility decreases with increase of temperature, among other salts in the seawater is the main reason for scale formation.

Let us now look very quickly at the five distillation based processes. The first, the vertical tube distillation, is schematically illustrated in Figure 1. After appropriate treatment, the sea water is conducted through vertical metal tube bundles surrounded by steam to effect a heat exchange. As a consequence, the steam is condensed while converting part of the seawater to steam. In order to improve the efficiency, this is repeated in several chambers held at progressively lower pressures so that boiling can take place at correspondingly reduced temperatures, in this manner making possible the recovery of virtually all heat provided in the first chamber. The first chamber steam is provided by a steam generator plant.

The second distillation process, seen schematically in Figure 2, as in the preceding utilizes the fact that water boils at lower temperatures at lower pressures. The heated seawater is brought into a chamber of pressure sufficiently low to cause "flash" evaporation. This evaporation leads to lowering of the temperature in the remaining brine. Just as in the first distillation process, several chambers are used here to improve the efficiency of the heat exchanger process. This process has also been designated as single effect multistage or SEMS. An effect is a complete distillation step.

The third of the distillation processes is the so-called MEMS or multieffect multistage process. It overcomes the 4°F. per stage limitation of the SEMS and makes possible the addition of more stages for each temperature level. Here there is also the opportunity of using the highly concentrated brine for by-product recovery. Schematically it is shown in Figure 3. The utilization of a number of effects, each operating at different temperatures, also greatly improves scale control.

The fourth process, the vapor compression distillation, is illustrated in Figure 4 and is based on the fact that as vapor is compressed, its temperature and pressure increase while its volume decreases. The vapor formed in the special chamber of the first effect supplies heat to the seawater being pumped through the bundles of the second effect. As the vapor loses heat to the brine, it condenses, falling to the bottom as the product water is taken away.

This schematic of the two effect unit shows the essence of this approach with the primary difference from other distillation processes being the manner by which heat is introduced into the system, i.e., here mechanical work is converted into heat of compression. The bulk of the energy here is supplied by the motor which drives the compressor.

The last of the distillation-based approaches is the so-called solar humidification process. This process is based on the fact that water evaporates from free surfaces at temperatures considerably below its boiling point. The evaporation rate here is dependent on the water temperature and the relative humidity of the space above the water. This process is usually carried out in a solar still, and is schematically illustrated in Figure 5.

While the solar distillation approach seemingly offers considerable economic advantage because of its essentially free energy source, it actually is far from economical as there are severe limitations in output. One pint of water is the maximum output per square foot of sun energy absorbing surface. Though it is an interesting concept, it certainly is no contender for large scale desalination operations.

FIGURE 1: SCHEMATIC OF A VERTICAL TUBE DISTILLATION PROCESS

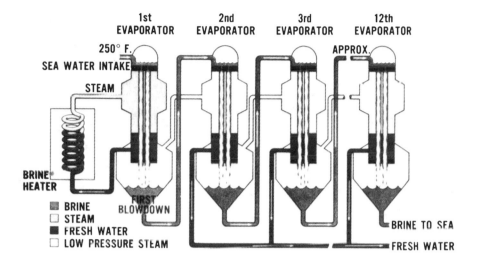

FIGURE 2: SCHEMATIC OF MULTISTAGE FLASH DISTILLATION PROCESS

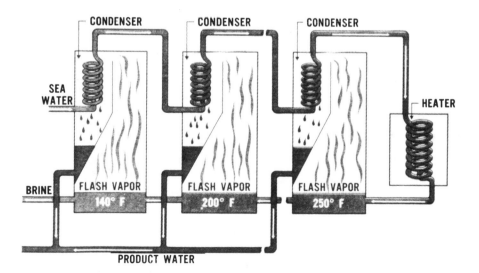

FIGURE 3: SCHEMATIC OF MULTIEFFECT MULTISTAGE DISTILLATION PROCESS

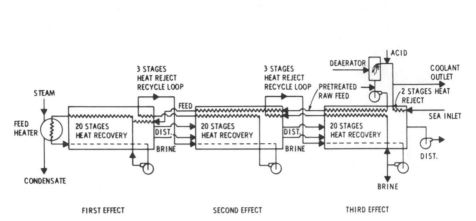

FIGURE 4: SCHEMATIC OF VAPOR COMPRESSION DISTILLATION PROCESS

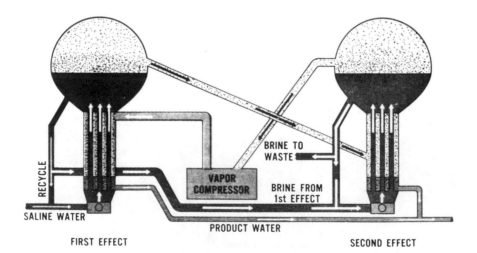

FIGURE 5: SCHEMATIC OF A SINGLE SOLAR STILL

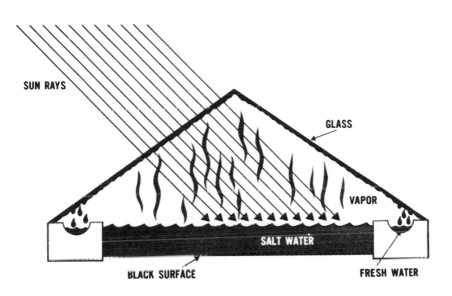

MEMBRANE PROCESSES

The next family of desalination processes are grouped under the title of membrane processes. While the three to be discussed vary notably, they all utilize
porous membranes permitting some substances to pass freely under given conditions while completely blocking passage of others. In two of the three, we
are concerned with ion migration under applied current and are essentially operating under the laws of electrochemistry.

The process in this group which is most advanced in principle is the so-called
electrodialysis process. It utilizes an electrolytic cell with two different ion
elective membranes (see Figure 6), one each permitting passage of anions and
cations. The imposed current drives the ions through the membranes leaving
fresh water between them. In addition to the membrane costs, the cost of
such processing is directly related to the salt content of the water, with current
costs at present making it uneconomical for salt water conversion.

Efforts in progress to operate such cells at elevated temperature where the electrical resistance of the electrolyte (seawater) is reduced show promise for the
future. Work also continues on membrane development, which also can have
important effects on the economics.

FIGURE 6: SCHEMATIC OF AN ELECTRODIALYSIS CELL

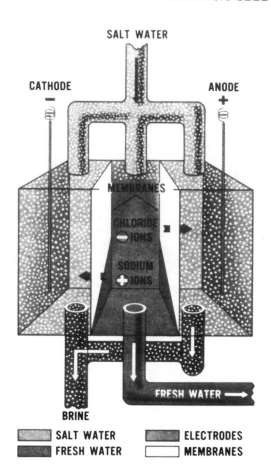

SALT WATER

CATHODE —   ANODE +

MEMBRANES

CHLORIDE IONS

SODIUM IONS

FRESH WATER →

BRINE

	SALT WATER		ELECTRODES
	FRESH WATER		MEMBRANES

The second of the membrane processes, the so-called transport depletion process, is based on the fact that there is a difference in the ion transport numbers in bulk solution and within an ion selective membrane. While in a simple electrolytic cell the sodium ion carries 40% of the current with the balance being carried by the chloride, by contrast, in a cation permeable membrane, the sodium will carry just about all of the current.

The principle of this process is schematically shown in Figure 7. Here the seawater enters the top of the cell and passes into the compartment separated by a nonselective membrane each bounded by a cation permeable membrane. Current drives the cations through the cation selective membrane while the anions pass through the nonselective membrane, depleting one of the compartments of salt (product) while enriching the others, both product water and concentrated brine being removed from the cell. Here, as in the electrodialysis process, progress is intimately tied up to membrane development. While the transport depletion process requires somewhat greater amounts of power, its use of only half as many membranes more than offsets the power cost differential in comparison with electrodialysis.

FIGURE 7: SCHEMATIC OF A TRANSPORT DEPLETION CELL

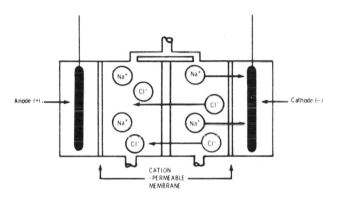

Reverse osmosis is the third of the membrane processes. It is based on osmosis, the phenomenon which describes the behavior of pure water and a salt solution on opposite sides of a semipermeable membrane. Here the pure water diffuses through the membrane and dilutes the salt solution. The water behaves as though it were under the effect of pressure, with the pressure being referred to as osmotic pressure. The osmotic pressure is a function of the differences in salt concentration and the temperature of the solution. By applying pressure on the salt solution, one can reverse the process. A schematic of the principle is illustrated in Figure 8, while the essentials of the process are shown in Figure 9.

In practice, reverse osmosis plants use several basic designs in order to effect desalination. The designs are a product of the objectives, namely, the need to use as large membrane surface areas as possible and also sustain the high pressures needed, which can be as high as 1,500 psi.

The schematics of plate and frame, spiral wound, tubular and hollow fiber approach are illustrated in Figures 10a through 10d, all of which are to be found in operational systems. Almost without exception, these systems are utilized in processing of brackish water not exceeding 3,000 ppm of salt content.

FIGURE 8: SCHEMATIC OF PRINCIPLE OF REVERSE OSMOSIS

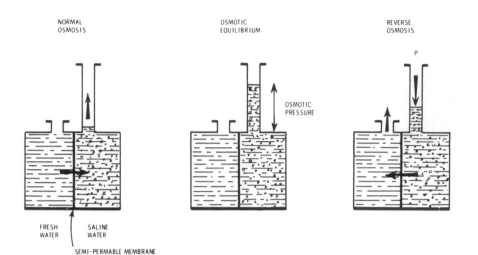

FIGURE 9: SCHEMATIC OF REVERSE OSMOSIS PROCESS

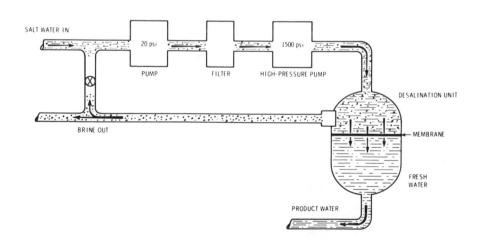

FIGURE 10: FOUR COMPETING REVERSE OSMOSIS APPROACHES

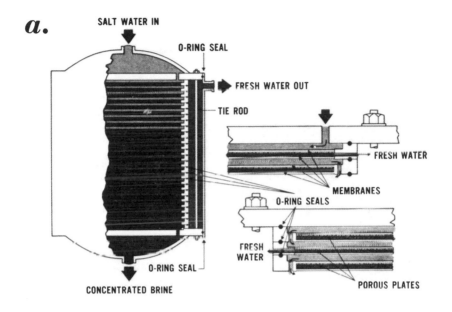

a.

Plate and Frame

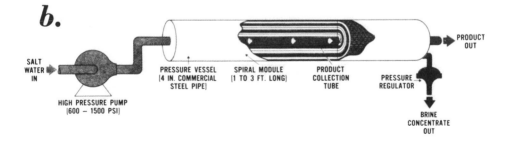

b.

Spiral Wound Membrane

FIGURE 10: (continued)

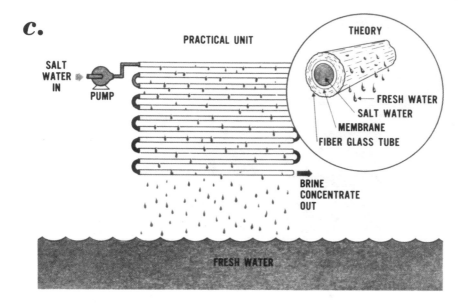

Tubular

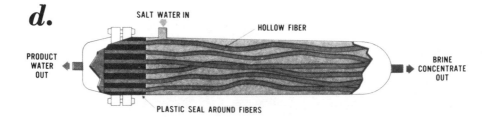

Hollow Fiber

Seawater is the most exciting potential market for the reverse osmosis process in particular, and for all the membrane processes in general, with realization of this potential depending on the success of membrane and related research in progress. The low power consumption, the low operating temperature and corresponding reduction of corrosion problems all favor the membrane processes.

CRYSTALLIZATION

The third important group of desalination processes is based on the fact that on cooling to its freezing point a salt solution will deposit crystallites of pure water in some cases, and in others crystallites of pure water combined with special reagents. This approach in principle has some notable advantages over the other phase change approaches, i.e., distillation, because of two facts. First, at low temperatures corrosion problems are importantly reduced and, second, of the desalination processes which involve phase changes, the freezing process has the smallest energy requirements.

Two freezing-based desalination processes have been developed. The vacuum freezing vapor compression approach releases the latent heat of fusion when precooled water is introduced into a chamber at a low pressure. It is schematically illustrated in Figure 11.

FIGURE 11: SCHEMATIC OF VACUUM FREEZING VAPOR COMPRESSION
PROCESS

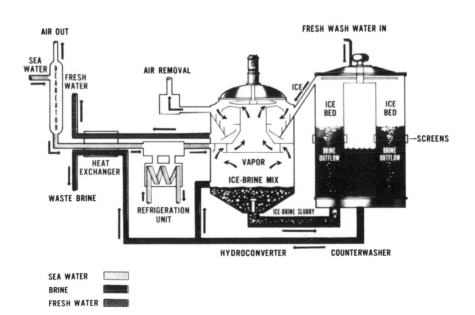

After deaeration and precooling by heat exchange with the product water and waste brine, the cold seawater is introduced into the lower chamber of the hydroconverter maintained at 3 Torr. As a consequence, almost a seventh of the water evaporates removing heat from the seawater and freezing almost half of it. The ice water slurry is gradually pumped into the counterwasher where it is washed and eventually pushed back to the upper section of the hydroconverter. The water vapor formed in the hydroconverter in the flashing of the seawater is compressed by a special unit located at the top of the hydroconverter. The heat contained in this vapor is that removed from the seawater, and it goes to melt the ice into fresh water.

The alternate process, the secondary refrigerant freezing process, differs from the first in that the freezing is accomplished by directly evaporating a reagent such as butane into the water. This process, schematically illustrated in Figure 12, also uses counterflow rinsing of the frozen ice crystallites.

FIGURE 12: SCHEMATIC OF SECONDARY REFRIGERANT FREEZING

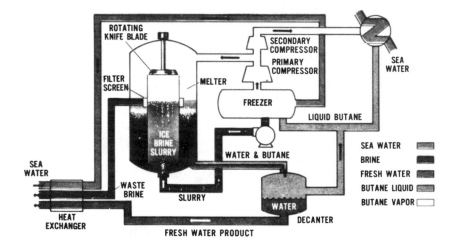

The butane vapor which contains the heat removed from the ice is compressed in the primary compressor and then introduced to the melter, condensing upon the ice, and in the process giving up heat and melting the ice. The liquid butane and the product water flow to a decanter where they are separated.

The hydrate formation process is an example of desalination through crystallization of water combined with an alternate substance. Here seawater and a hydrating agent such as propane, for example, are brought together in a crystal-lizer under suitable temperature and pressure conditions to form hydrate crystals. While as in the preceding processes the crystals do not contain salt, they contain water and propane. This type of crystal is referred to as solid clathrate crystal (see Figure 13). Very much as in the two processes just discussed, the crystals are separated from the brine and washed with small amounts of water. Subsequent to the separation, propane is condensed on the crystals, melting them.

FIGURE 13: DIAGRAM OF A CLATHRATE HYDRATE CRYSTAL

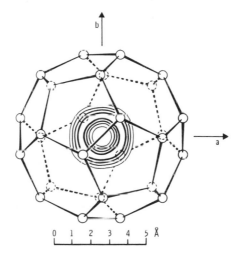

The schematic of the process is not much different from that given in Figure 11. On melting of the crystals, one has two immiscible liquids, the product water going to storage tanks and the propane being returned to continue the desalination processing. The economic feasibility of this process is far from established and work is being done in developing better hydrating agents.

CHEMICAL PROCESSES

The last potentially viable desalination approach is the chemical approach, or specifically the ion exchange process. Desalination here is accomplished by providing resins (reagents) which interact either with the salt or the water itself to form compounds which are readily separated. An ion exchanger is a porous bed of certain resin materials that have the ability to exchange ions in the resins with the solution which is in contact with the resin. In practice, both cationic and anionic ion exchangers are used. Most often they are placed in series, although mixed beds can also be used. In Figure 14 a cation exchange replaces the sodium with hydrogen ion, converting the dilute salt solution to dilute hydrochloric acid solution. When the acid solution passes through the anion exchanger, the chloride ion is exchanged with the hydroxide ion. The hydrogen and the hydroxide ions combine to form pure water molecules.

As the process continues, the resins are saturated until they completely lose the ability to remove either sodium or chloride ions. At this time the resin must be regenerated, a process which is usually accomplished by washing the resin beds with acids and bases to restore the original ion exchange properties. Regeneration is costly and for most practical purposes ion exchange application is limited to situations where salts are in low concentration and below 3,000 ppm.

FIGURE 14: ION EXCHANGE PROCESS

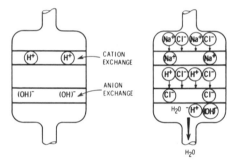

The preceding pages have attempted to present a bird's eye view of the principles involved in desalination and the many schemes utilizing these principles for desalination operations. The papers to follow will take the more important approaches and discuss them in more detail.

ACKNOWLEDGEMENT

The writer wishes to state that this write-up is based on the Office of Saline Water's publication The ABC of Desalination. The illustrations used here also are from the same source. The writer wishes to thank the OSW for permission to reproduce stated figures.

DESALINATION — CHALLENGE AND OPPORTUNITY

J.W. O'Meara, Director
Paul C. Scott
Office of Saline Water
U.S. Department of the Interior

ABSTRACT

The history of the U.S. Government's desalination program is briefly reviewed and the current U.S. test facilities are described. Recent advances in distillation, membrane processes and crystallization processes are summarized. Plans for future desalting efforts are presented. It is concluded that development of a number of different desalination processes are needed, since each potential location for a desalination plant has its own special characteristics and problems.

INTRODUCTION

The possibility of obtaining fresh water from saline water has been known to mankind for more than two milleniums, but the need for desalination has only arisen in recent times as a result of several factors. These include rising population, improved standard of living, and increasing industrial development. In the United States it is estimated that within the next two decades no additional conventional new sources of fresh water will be available in such regions as the upper Rio Grande-Pecos River, the Colorado River, or the Upper Missouri River.

In recognition of desalination as one possible solution to the water problem, the U.S. Congress, in 1952, passed the first Saline Water Conversion Act establishing the Office of Saline Water and providing $2 million for development of "practical means for the economical production from sea and other saline waters of water suitable for municipal, industrial, agricultural and other uses." For the first five years the Office of Saline Water was a modest operation, with a couple of dozen employees and average annual appropriations between one-half and one million dollars. Emphasis was on laboratory and small pilot plant feasibility studies. In 1958, under the provisions of legislation that was introduced by

19

Senator Clinton P. Anderson, the OSW was authorized $10 million to build five desalting plants to demonstrate the feasibility and potential of different desalting processes in different regions of the country.

In 1961, the Anderson-Aspinall Act (P.L. 87-295) was passed earmarking $75 million for research and development during the fiscal years 1962 to 1967, and giving the saline water program most of its current operating authority. In 1965 (P.L. 89-118), and again in 1971 (P.L. 92-60), the Congress expanded and accelerated the Saline Water Conversion Program and recent OSW budgets have ranged between $25 and $30 million per year.

CURRENT OSW ACTIVITIES

Facilities

During the past development period, large test facilities have been constructed and tests conducted to prove out desalting technology which was originally initiated in the research laboratories. The OSW now has five sites at which the field test development program is conducted. The locations of these sites and their general equipment are summarized in Figure 1.

FIGURE 1: LOCATION OF MAJOR OSW TEST FACILITIES

FIGURE 2: WRIGHTSVILLE BEACH TEST FACILITY

The Wrightsville Beach Test Facility, shown in Figure 2, is used specifically to test new seawater conversion processes, techniques, and materials at the pilot plant level. There are a number of pilot plants at this location varying in capacity up to 60,000 gallons per day. The test programs on twenty pilot plants have been completed and the plants have been either dismantled or transferred to other sites for additional testing. The plants are primarily involved in developing and testing new techniques or materials that can be incorporated into future higher-efficiency plants, for instance, slurry seeding for calcium sulfate scale control, or use of aluminum and aluminum alloys.

Originally a one million gallon per day vapor compression distillation (VC) test bed plant was built at the Roswell, New Mexico, test facility, shown in Figure 3. The plant was plagued with repeated breakdowns of the unprecedentedly large mechanical vapor compressor and was shut down for reevaluation of the process. In 1969, major modifications to the VC test bed plant were initiated, including the recent replacement of the mechanical vapor compressor with a high temperature water jet thermocompressor currently under test.

The Roswell site also includes a brackish water test center that provides facilities for the testing of a number of reverse osmosis and electrodialysis membrane pilot plants on ten synthetically blended brackish waters most common in the United States. Also at this facility OSW evaluates pumps and other apparatus to be used in conjunction with the reverse osmosis process.

FIGURE 3: ROSWELL TEST FACILITY

FIGURE 4: FREEPORT TEST FACILITY — 1967

At Freeport, Texas, the one million gallon per day vertical tube evaporator (VTE) test bed plant, shown in Figure 4, has completed the programmed development tests and demonstrated significant advances in both economics and process reliability. This plant has now been modified to incorporate advanced features of the multistage flash (MSF) process. In fact, this plant now has combined, in modular-type construction, the falling-film evaporator, enhanced tube surface, and concrete-lined vessel features of the VTE process with MSF preheating.

It is anticipated that combining these features will result in a significant decrease in capital investment for plants of 50 million gpd or larger. This will be described in greater detail later.

Webster, South Dakota, was for many years the site of a 250,000 gpd electrodialysis plant. The operation of this plant was discontinued in June 1971 since the plant had served its purpose. The test facility is continuing in operation at Webster for evaluation of pretreatment and membrane systems.

The last of the five test facilities is the San Diego Test Facility at which a number of plants and pilot plants for seawater distillation are currently under test. Among these are the Clair Engle Desalting Plant shown in Figure 5.

This is a one million gallon per day multieffect, multistage (MEMS) distillation plant. The special effect (E-1A), which in conjunction with the lime magnesium carbonate (LMC) pretreatment plant, permits test operations of the test bed plant up to approximately 335°F., is also at San Diego.

FIGURE 5: CLAIR ENGLE DESALTING PLANT

San Diego is also the location of the VTEX, a test vehicle used for studying the performance of large tube bundles in the VTE process. In addition, another test vehicle, the HTE-X, is being constructed at San Diego in order to carry out a major study of the performance of large horizontal tube bundles in multiple effect distillation processes.

Processes

Major efforts at the present time are directed toward three basic methods of desalting. These are distillation, membrane and crystallization (or freezing) processes.

Of these three, the first, distillation, has seen the most employment in commercial plants and has reached the point of large scale development of components. One example is the module at San Diego Test Facility, shown in Figure 6. This module, based on the multistage flash process, has an output of two and a half million gallons per day and has been in operation for three years. It is the data derived from this plant that is being used for planning plants up to 50 mgd in capacity.

FIGURE 6: MSF MODULE AT SAN DIEGO TEST FACILITY

Figure 7 shows the general concept of the module in relation to plant. Although it has an output of only two and one half million gallons per day, the recirculation pump and flow quantities are those which would be used for a one-third section of a 50 million gallons per day plant. By operation of this module at both low and high temperatures, complete performance of a much larger plant can be simulated.

FIGURE 7: GENERAL ARRANGEMENT — TEST MODULE

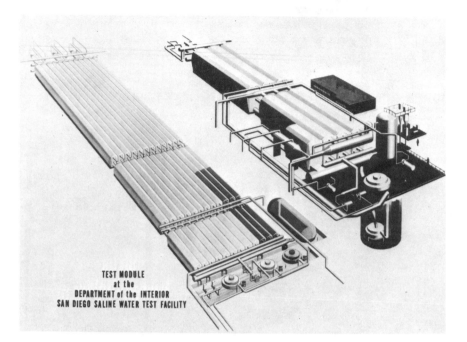

TEST MODULE
at the
DEPARTMENT of the INTERIOR
SAN DIEGO SALINE WATER TEST FACILITY

Although multistage flash distillation is the primary desalting process used to-day, engineering and development effort by the Office of Saline Water has in-dicated that it may be possible to combine this process with the vertical tube evaporation process to significantly reduce the cost of product water. Such a combined process is illustrated in Figure 8.

The upper section of the plant is the vertical tube evaporation section in which seawater passes down the inside of large tubes and is evaporated by high tem-perature water vapor circulating in these tubes. The steam thus produced acts as the means of evaporating additional water in the next effect, while at the same time being condensed to yield fresh water. A number of vertical effects can be used, each operating at a lower pressure.

The multistage flash portion of the process is shown below the vertical effects and essentially acts as a feed water heater for the vertical tube section, although fresh water is produced in the multistage flash portion of the plant. As noted, salt water is pumped upward from the multistage flash stage to feed the vertical effects. Of the total production, about 85% of the fresh water is produced in the vertical tube section and 15% in the multistage flash section.

FIGURE 8: FLOW DIAGRAM OF THE VTE-MSF PROCESS

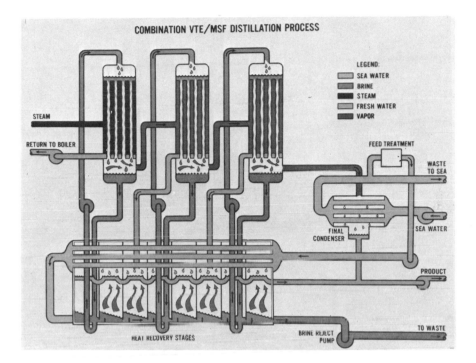

COMBINATION VTE/MSF DISTILLATION PROCESS

As mentioned before, OSW has recently placed in operation at Freeport a test plant of approximately one million gallons per day capacity employing this principle. This plant is shown in Figure 9.

Additionally, OSW has just begun construction of a 3 million gallons per day segment of a 12.5 million gallons per day plant employing this combined process. This module is located south of Los Angeles, in Orange County, California. It is designed to produce fresh water to maintain the high standards of Orange County's groundwater and to prevent a threatened incursion of salinity from the sea. The module could be expanded to produce 12.5 million gpd, which would make it the largest single plant of its kind in the world.

In essence, the OSW-Orange County Water district effort meets both the needs of the nation in terms of long-term technology development and the needs of Orange County in terms of short-range water requirements. While the module is producing water for Orange County, it will also be producing engineering data to help to advance desalting technology.

FIGURE 9: FREEPORT TEST FACILITY — 1971

To develop data on the flow and heat transfer characteristics of large vertical tube bundles such as used in this module, the Office of Saline Water has been testing such bundles. The test vehicle is shown in Figure 10 and contains two large bundles, each forming a 10 foot cube.

Also in the area of distillation, the Office of Saline Water has under development a combined process employing a gas turbine vapor compressor in conjunction with both vertical tube evaporation and multistage flash. The flow diagram for this plant is shown in Figure 11.

The plant size being studied at this time has a capacity of 8 million gallons per day and the vapor compressor is under construction. This unit would have application to areas where cheap natural gas is available and will result in a lower water cost than that achieved by single processes in this size plant. In this case, the vapor compressor acts as the energy source for the vertical tube evaporation section.

FIGURE 10: VTE TEST VEHICLE

FIGURE 11:   FLOW DIAGRAM OF VC-VTE-MSF PROCESS

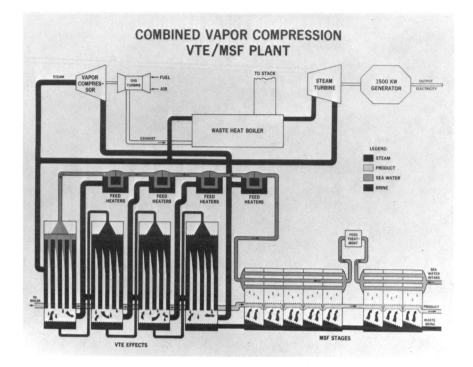

Turning now to membrane processes, the Office of Saline Water has had under investigation two membrane desalting systems applicable to brackish water. The development effort on the first of these, electrodialysis, has essentially been completed except for a program in the area of high temperature operation.

The major thrust in the membranes area at the present time is in the development of reverse osmosis systems, both for brackish and seawater desalting. In the brackish water area, large membrane assemblies have been developed with capacities up to 35,000 gallons per day. A number of these can be interconnected to form a large plant. Some idea of the size of a 250,000 gallon per day plant can be obtained from Figure 12. Additionally, due to the desire to obtain as much fresh water as possible from the feed water entering a reverse osmosis plant, efforts are now underway to design and construct a 695,000 gallons per day plant which will be capable of recovering 95% of the entering feed water as fresh water.

A number of companies in the United States are now involved in the production of reverse osmosis units which can be applied to the desalting of brackish water. The primary configurations used for brackish water desalting are tubular, spiral wound, and hollow fine fibers. The hollow fine fibers are a recent development and have the advantage of providing a large area of membrane surface to be packed into a small volume. The fibers are about one-third the diameter of a human hair and are produced by conventional fiber production methods.

FIGURE 12: GENERAL ARRANGEMENT — 250,000 GALLONS PER DAY REVERSE OSMOSIS PLANT

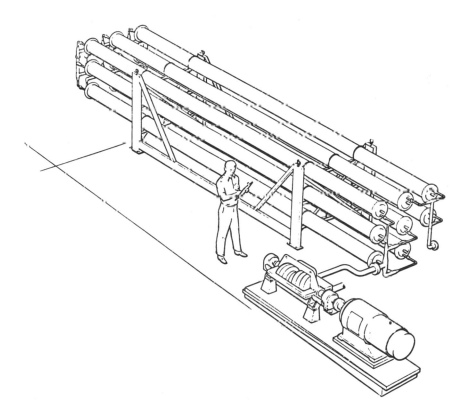

Due to the progress that has been made in membrane development, it now appears that reverse osmosis can be used effectively for the desalting of seawater. Two approaches are possible. The first is the use of two stages to reduce the salinity of the seawater from approximately 35,000 parts per million to approximately 3,000 parts per million in the first stage, and then pass it to a second stage, where it is desalted to a potable level of less than 500 parts per

million. The second approach is single stage desalting of seawater in which
the seawater is reduced from 35,000 parts per million to a potable level in
one stage. In the latter case a membrane is required with salt rejection of
over 99% in order to have a potable product. It also requires that the membrane operate at a high pressure, of the order of 1,000 to 1,500 pounds per
square inch.

With membranes which are now available, we feel that a two-stage seawater
desalting plant could now be built. It is foreseen, however, that with the progress made in our laboratories, single-stage seawater desalting by means of reverse osmosis will be a reality within the next few years.

In other work on membrane processes, pretreatment investigations are being
conducted and an extensive development program is underway to improve
membrane flux and life while achieving low cost. For example, it is now possible to have a packaged membrane system having a diameter of 8 inches and
approximately 30 feet long that will have a product water output of approximately 35,000 gallons per day. This employs the spiral wound unit. In the
hollow fine fiber area, membrane modules 8 inches in diameter and approximately 4 feet in length have an output of 3,500 gallons per day.

It is foreseen that developments of the membranes themselves may further increase the capacity of these units. Additionally, membrane life is now projected at 3 to 5 years by manufacturers and the Office of Saline Water has had
a variety of high performance membranes on test for over a year and a half on
actual brackish waters.

The technological "spin-off" from membrane processes is providing benefits to
the public in such areas as medicine, food production, wastewater renovation,
and the recovery of useable by-products from wastes.

In the cheese-making industry, for instance, about 70% of the 1.25 billion
pounds of whey produced annually is wasted and overburdens sewage treatment facilities or causes pollution by direct discharge into surface streams
and lakes. Whey is a high concentration of nutrients. Reverse osmosis and
electrodialysis can be used for concentrating the constituents of whey and
permitting their recovery either as high grade protein or as milk sugar.

In the wood pulping and paper production processes large quantities of water-
borne waste materials are generated. It is possible to concentrate the spent
liquor from the digestion of wood pulp by evaporation and recover the pulping chemicals and organic materials by electrodialysis. Thus, one can obtain
a reusable by-product and tend to purify water leaving the plant as waste.

The versatile RO process is also being utilized to concentrate fruit juices and
coffee, to remove various viruses and bacteria from contaminated liquids, and
is being developed for space application whereby liquid body wastes can be
converted to a potable product.

Although the Office of Saline Water has been involved in work on crystallization, or freezing, during the past ten years, new developments have indicated that this process should be reemphasized. In the past, the Office of Saline Water has supported development of the vacuum freezing vapor compression system in which ice is produced by lowering the pressure above a chilled salt solution. This ice is then washed and melted to produce fresh water. The compressor required to maintain the vacuum freezing process operation is essentially size limited, and therefore very large freezing plants using single compressors did not appear economically feasible.

The Office of Saline Water did operate a freezing plant with a capacity of over 100,000 gallons per day using this principle, but due to the compressor limitation, it was not felt that the normal reduction in product water cost could be obtained by increasing the size of components. Consequently, investigations are now underway on the use of ejectors to maintain vacuum and these ejectors are less subject to size limitation. Bench-scale developmental efforts are underway on the ejector method of maintaining vacuum.

Another freezing process under development employs a refrigerant which, when mixed with chilled seawater, will result in the formation of fresh ice crystals. These are then washed and melted to produce fresh water and the refrigerant is recovered for reuse in the process. Such a process is also not limited in size relative to plant application. Based on these laboratory and bench-scale efforts, it is planned that within the next year pilot plants will be constructed to establish complete systems performance.

Associated with all processes are analytical studies, laboratory investigations, and field tests of components so as to improve system performance. In the field of distillation these efforts include evaluation of pretreatment systems to decrease the possibility of the formation of scale on the heat transfer surface and improvement of heat transfer, both in multistage flash and vertical tube evaporation processes. In the case of the latter process, the heat transfer coefficients have been more than doubled for the same tube diameter by means of fluting of the tubes. An example of this is shown in Figure 13.

Materials development has also been emphasized and extensive work has been done on the application of aluminum alloys to distillation plants. Results so far indicate that these alloys could be used if proper design consideration is given to use of other materials in the same plant. For example, the combination of aluminum with steel without proper insulation results in galvanic corrosion which tends to destroy the aluminum. Consequently, the designer has to be careful in how dissimilar materials are used in a plant.

A relatively new area of OSW interest is the desalination of geothermal brines. The Department of the Interior has formed a task force which will utilize the best expertise of the bureaus and agencies of the Department for an in-depth coordinated stury of the potential development, for both power and water, of the geothermal resources of the Imperial Valley region of California. An important part of this task force effort will be an evaluation of the potential of existing processes to desalt mineral-laden geothermal brines and to develop new processes or combinations of processes, as required, to recover economically fresh water from this presently untapped source of supply.

FIGURE 13: SECTION OF FLUTED TUBE FOR VERTICAL TUBE EVAPORATORS

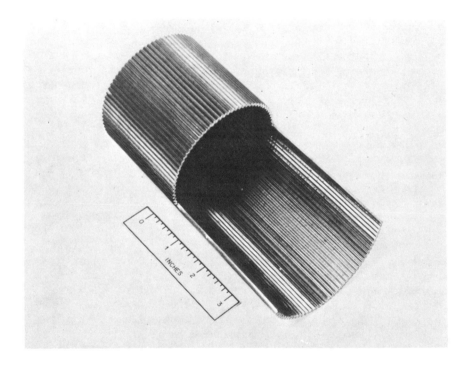

The economical recovery of water from geothermal brines differs in several respects from normal desalination. First of all the brines are initially hot, with temperatures of 400°F. and higher. These brines are corrosive, and often laden with suspended solids. They are usually saturated with silica which can form scale, and hydrogen sulfide and carbon dioxide which are corrosive gases. There is a problem of disposing of the brine without damage to the environment.

To investigate and overcome these problems, the OSW has undertaken the design and construction of a 50,000 gpd pilot plant to be operated in the Imperial Valley during 1972, as well as associated bench-scale studies on the problems of silica deposition, dissolved gases and corrosion in geothermal brine systems. The long-range objective of all this work is the development of a geothermal resouce to supply usable water for municipal and industrial development in this area of

the country, and to investigate the opportunities for augmentation of the
Colorado River basin's water supply.

One of the OSW program goals is the construction of first-of-a-kind production
or prototype plants to desalt both sea and brackish waters. The purpose of the
prototype plant program is to overcome the major obstacles to introduction of
new technology, namely the costs and risks associated with doing what has not
been done before. Until a number of years of operating experience have been
obtained under a variety of conditions, the first users bear costs and acquire
knowledge which permit subsequent users to reduce costs and improve effectiveness. If uncompensated, the pioneers subsidize the latecomers. Worse yet, the
pioneering efforts simply do not take place or take place too late to be effective.

Few public water supply agencies have any experience whatsoever in water manufacturing by desalting; hence, there is an obvious reluctance to adopt the new
technologies even when analysis would indicate major savings. These risk, experience and knowledge factors all suggest that additional incentives are required
if desalting technology is to be advanced to a viable water manufacturing industry in a timely, cost-effective way. The incentives may consist of technical assistance, financial assistance, or both.

Thus, the OSW is concentrating its efforts toward the development of various
types of prototype plants. Under consideration are the following:

(1)   a 30 to 50 mgd seawater conversion prototype plant in
       California,
(2)   an 8 mgd distillation plant in Texas as well as other proto-
       types in the southwestern United States,
(3)   a 3 mgd membrane plant at Foss Reservoir in Oklahoma.

In all cases, the major objective of the Office of Saline Water is to develop the
technology so that fresh water costs from desalting plants will be more competitive with other means of obtaining water. In some cases, desalting may be the
only method to obtain new water supplies of acceptable quality.

The OSW program has evolved, during the last twenty years, from a small ex-
ploratory and testing program to one involving or contemplating desalination
plants in the size range of tens of millions of gallons. In the past five years the
budget of the Office of Saline Water has ranged between $25 and $30 million.
It is foreseen that with increased applications and the necessary supporting effort for these applications, the budget will increase in the coming years.

As of this date, there are an estimated 700 desalting plants of a capacity of
25,000 gallons per day or larger throughout the world. These plants are capable
of producing an approximate total of 350 mgd of fresh water. The best reported cost of water at a U.S. plant is 35 cents per 1,000 gallons for a 1.2 mgd
electrodialysis plant (Siesta Key, Florida) using as feed water a brackish water
containing approximately 1,400 ppm total dissolved solids. For seawater, the

reported cost is 95 cents per 1,000 gallons for a 2.6 mgd multistage flash distillation plant at Key West using seawater. Both of these figures are believed to represent actual costs on the basis of 1970 dollars, 90% load factor and 7% fixed charges.

A dual purpose power and water distillation plant at Rosarito, Mexico, with a capacity of 7.5 mgd, reports a cost of 65 cents per thousand gallons for seawater desalination.

These examples illustrate and stress the need for a number of different processes depending upon the particular problem situation. Developing a desalting technology is not a simple matter of developing the one "best" process. Different applications are optimum in certain situations and each has its own special problems to overcome whether it be the concentration of salts in the saline feed water, the types of ions present, the intermittency with which the plant will operate, the economic life required, and the type, amounts and costs of energy available.

ALUMINUM FOR DESALINATION SERVICE

Ellis D. Verink, Jr.
University of Florida

ABSTRACT

Based on service experience and on the results of pilot plant tests, aluminum alloys have been shown to be suitable for the construction of desalination plant equipment. The suitability of aluminum for this service makes available to de signers the great economy implicit in aluminum alloy construction and should contribute importantly to the decrease in cost of water. Results of pilot plant operations are reviewed. A list of references of recent literature on use of aluminum alloys in desalination apparatus is included.

USE OF ALUMINUM ALLOYS

The widespread use of aluminum alloys in the chemical, petroleum and marine industries supports the contention that the aluminum alloys have a strong and vital place in the construction of desalination apparatus. Several test plants have been constructed to demonstrate the utility of aluminum alloys for the construction of desalination apparatus.

One of these, operated by Reynolds Metals Company, has been in operation at the OSW Test Station in Wrightsville, Beach, North Carolina, for over two years (1). Another has been operated by Dow Chemical Company for the Aluminum Association Task Force on Desalination. This unit is located at the Materials Test Center of the Office of Saline Water at Freeport, Texas (2)-(5). An aluminum alloy flash distillation plant has operated in France for three years (6), and a desalination plant constructed largely of aluminum has given excellent service in Israel (7).

The establishment of aluminum alloys for desalination service is consistent

with the desire to reduce the cost of desalted water. Maximizing of economy
in the production of potable water from desalination plants requires the use of
lowest cost materials having the requisite performance reliability. Considering
the large portion of the capital cost involved in the purchase of heat exchanger
tubing for flash desalination plants, a change from copper-base alloys to aluminum alloys for heat exchanger tubing could provide a significant decrease in
initial capital costs, which would result in continuing savings in debt-service, etc.

Although heat exchanger tubes for desalination represents a tremendous potential market for aluminum alloys, tubing is by no means the only semifabricated
form of aluminum useful to this industry. The outstanding resistance of aluminum alloys to weathering in marine atmospheres recommends their use for other
services where maintenance costs or appearances are important. Such applications include supporting structures, weatherproofing for thermal insulation, instruments for airlines, electrical wiring and conduit, handrails, grating and stair
treads.

Of particular importance is the test program jointly conducted by The Aluminum Association, Dow Chemical Company and the Office of Saline Water at
Freeport, Texas. Figure 1 shows a flow chart for this test unit. The concept
of this pilot plant involves the simulation wherever possible of design conditions
for the Pt. Loma No. 2 MEMS Plant.

FIGURE 1: THE ALUMINUM ASSOCIATION'S ALUMINUM SEAWATER
DISTILLATION UNIT

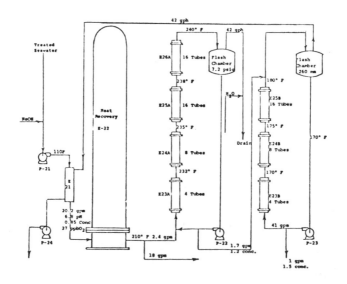

Average operating conditions from February 13 to September 8, 1970.

The simulation included matching temperature ranges, brine concentrations, alloy-surface-to-flow-volume ratio, blending of process streams, heat transfer conditions and "residence time." The alloys tested included four heat exchanger tube alloys, two plate alloys and a pipe alloy (see Table 1). Flow velocities were 2.5, 5.0 and 10 feet/second. Brine concentration ratios ranged from unconcentrated brine to approximately 2.25 concentration. The temperature ranges were as follows: in the Preheater Section, 85° to 209° F.; in the First Effect, between 225° and 235° F.; the Brine Heater, between 235° and 250° F.; and the second Effect, between 160° and 180° F.

TABLE 1: COMPOSITION AND MECHANICAL PROPERTIES OF ALUMINUM ALLOYS

Alloy	Minimum ksi		Nominal Composition, Percent					
	T.S.	Y.S.	Cu	Mg	Mn	Si	Cr	Zn
3003-H14	20	17	0.12		1.2			
5052-H34	34	26		2.50			0.25	
5454-H34	39	29		2.70	0.8		0.10	
6061-T4	30	16	0.25	1.00		0.6	0.20	
6063-T832	41	36		0.70		0.4		
7072*								1.0

*Cladding on Alclad 3003

After two years of operation, The Aluminum Test Plant remained in excellent condition and had been in operation more than 99% of the time (5). The test results reaffirmed the earlier contention that the economics implicit in the use of aluminum alloys for desalination plant equipment are available to designers of process equipment. Aluminum piping, heat exchanger tubes and process equipment showed themselves to be highly resistant to the desalination plant exposures.

Pitting of heat exchanger tubes was experienced in two locations within the plant after one year of service. These locations were where the water first contacted the aluminum equipment in the Preheater Section (first pass), and in the Second Effect, where concentrated brine and very low water velocities (2.5 feet per second) resulted in deposition of solids in the tubes under conditions which stimulated some localized attack. However, the pitting of heat exchanger tubes which was observed after one year apparently had stopped since the depth of attack after two years was no greater than it had been after one year.

Galvanic corrosion can occur between dissimilar metals, such as stainless steel-to-aluminum joints (e.g., between pumps and piping), in the brine circulation system. By use of simple design practices, such as the insertion of short lengths of plastic lined pipe between the dissimilar metals (without electrical isolation), it is possible to reduce such dissimilar metal action to tolerable proportions. It is also possible to mitigate dissimilar metal problems by inclusion of sacrificial metal anodes (zinc or aluminum alloy) at flanged connections.

The economics of aluminum heat exchanger tubes is discussed in Reference 8. Figure 2, taken from Reference 8, shows the financial break-even times for aluminum versus copper-nickel tubes for equal capitalized costs for a given size of heat exchanger tube. Although the prices from which this calculation was made are now out of date, the results of the calculation are still qualitatively valid.

FIGURE 2: FINANCIAL BREAK-EVEN TIMES FOR ALUMINUM VERSUS COPPER-NICKEL TUBES FOR EQUAL CAPITALIZED COST

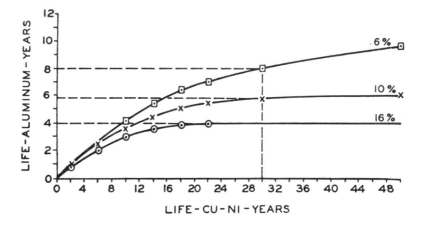

From Figure 2 it is seen that if aluminum tubes last six years, this would be equivalent to thirty years life for copper-nickel alloys for a return on investment of 10%, and that any longer life for aluminum would make aluminum less expensive than Cu-Ni, even if the Cu-Ni tubes lasted forever. The economics are even more favorable when other, higher cost alternatives are compared with aluminum. The excellent performance of aluminum in the test plant at Freeport, Texas, as well as in chemical, petroleum and marine exposures suggests that long life may be expected.

FIGURE 3: ALUMINUM ALLOY 5454 EMPLOYED FOR CONSTRUCTION OF FLASH CHAMBERS

This unit was operated at 7.2 psig and 240°F. No evidence of corrosion was found. The resistance of aluminum weatherproofing to the severe marine environment is evident.

FIGURE 4: GALVANIC CORROSION OF 6061 ALLOY WELD NECK FLANGES

This occurred at points of dissimilar metal contact with stainless steel exposed to brine. Use of a short length of plastic-lined pipe between the stainless steel and the aluminum pipe mitigates this problem without need for electrical isolation.

REFERENCES

(1)   U.S. Dept. of the Interior, "Operation of a Multistage Flash Distillation Pilot Plant Which Has Aluminum Heat-Transfer Surfaces," Saline Water Conversion Report 1969-1970, Office of Saline Water, pp. 358-370.

(2)   Verink, Jr., E.D., "Dynamic Testing of Aluminum Alloys for Desalination Service," Proc. of 25th NACE Conf., National Association of Corrosion Engineers, March 10-14, 1969, Houston, Texas, pp. 339-341.

(3)   Verink, Jr., E.D., "Construction and Operation of Dynamic Test Loop of Aluminum Alloys for Desalination Environments," Proc. of 26th NACE Conf., National Association of Corrosion Engineers, March 2-6, 1970, Philadelphia, Pa., pp. 597-599.

(4)   Verink, Jr., E.D., "Aluminum Alloys for Desalination Service — A Progress Report," CORROSION/71, Paper No. 60, Chicago, Ill., March, 1971.

(5)   Verink, Jr., E.D., "Performance of Aluminum Alloys in Desalination Service," CORROSION/72, Paper No. 27, St. Louis, Mo., March 20-24, 1972.

(6)   Vargel, C., "Aluminum Alloys in Sea Water Desalination Plants," Paper presented at the 3rd Internat'l. Symposium on Fresh Water from the Sea, Dubrovnik, Yugoslavia, Sept. 13-17, 1970.

(7)   Pachter, M., Barak, A. and Weinberg, J., "Reducing Costs by Using Aluminum Tubes in Multi-Effect Desalination Plants," Proc. of the 8th Nat'l. Symposium on Desalination, En Boqeq, Israel, March 14-15, 1971.

(8)   Wanderer, E.T. and Wei, M.W., "Applicability of Aluminum for Desalination Plants," Metals Engineering Quarterly, Vol. 7, No. 3 (August, 1967), pp. 30-36.

COPPER ALLOYS FOR DESALTING PLANTS

Arthur Cohen
Copper Development Association

ABSTRACT

The usefulness of copper alloys in seawater and desalting plant service is described in terms of their properties, benefits and limitations. A statistical approach for predicting tube failures is shown to be both necessary and reliable because of the large number of tubes involved.

The results of the corrosion behavior of 12 copper alloys after up to 27 months exposure to seawater under conditions simulating the brine heater effect of a multieffect, multistage flash desalting plant are described. Influence of tube manufacturing methods is also covered.

Comparison is made of the material performance in the various sections of OSW's million gallon per day desalting plant. Finally, a description of DCA's new desalting plant is given highlighting the operating conditions which simulate full size plants.

INTRODUCTION

Over the centuries copper, brass and bronze have probably been selected as the materials of choice to resist corrosive environments more than any other family of metals. It is very likely that there was once a time when all corrosion resistant applications were filled by copper and its alloys. Today, there are other families of materials that also resist corrosive environments, each according to its own special properties. And now copper and its alloys must compete for applications in the desalting industry with other more or less useful materials.

It would be unrealistic to expect to provide in one paper a complete guide to the use of copper and its alloys in the desalting field. A treatise by multiple authors engaged actively in the field would be required for that. This paper has more realistic and modest goals.

In the interest of sorting out the technical considerations involved in this materials discussion, an effort will be made to establish the basis for copper's usefulness by sorting out the various copper alloy systems, describing some of the properties, benefits and limitations provided by these copper alloy systems, and finally highlighting some of the studies that have been completed or are presently underway in the evaluation of copper-base materials for desalting plant service.

Along with heat transfer characteristics, the corrosion resistance of copper and its alloys is the basis for their usefulness in heat exchanger applications. But why are these alloys corrosion resistant? A metal that is corrosion resistant is either inherently "noble" by virtue of its low driving force towards an oxidized state, or else it is protected against attack by the presence of an adherent and impervious film of corrosion product on its surface. With copper, you can have it both ways. It is clearly a member of the nobility in the chemical sense and in addition to being a noble metal, copper also exhibits film-protected behavior. The environment determines which type of behavior is exhibited.

STANDARD ALLOYS

The list of standard compositions includes more than 200 wrought alloys and over 100 cast alloy compositions. For heat exchanger applications in desalting plants, the interest is primarily in tube and in only a relatively few of these alloys. But what are the special virtues of the specific alloy families useful in heat exchanger service? What are the pertinent facts of the alloying elements that are added to copper to improve its engineering properties (1), including cost effectiveness?

Zinc is the alloying element most frequently added to copper. As zinc content increases, a point is reached at about 16% zinc where copper alloys become subject to corrosion by dezincification with further increase in zinc content. Also, alloys high in zinc have decreased resistance to stress corrosion.

Tin, added to copper-zinc alloys — that is to brasses — significantly improves those alloys for heat exchanger service, by increasing their strength and by increasing resistance of dezincification. Copper Alloy 260, 70-30 copper-zinc, plus 1% of tin becomes the basis for Copper Alloys 443 to 445. Adding ¾% of tin to Copper Alloy 280 makes it Alloy 464 in the standard series. Or, to use some old familiar terms, cartridge brass plus 1% tin equals admiralty. Muntz metal plus ¾% tin is naval brass.

Admiralty and naval brass are further inhibited against dezincification by small additions of arsenic, antimony or phosphorus. In the same way, the naval brasses (Alloys 464 to 467) are made highly resistant to dezincification through small additions of these alloying elements. Alloy 687 (aluminum brass) is also inhibited with arsenic additions.

The effect of aluminum on copper alloys is to provide resistance to high velocities and also resistance to corrosion by both clean and polluted seawater.

Finally, the addition of nickel to copper, along with small additions of iron, has the effect of increasing resistance to velocity and impingement attack, resistance to stress corrosion compared to other alloy systems, and especially resistance to seawater corrosion.

GENERAL MATERIALS PHILOSOPHY

Returning now to the general materials philosophy as it applies to desalting plant construction, the following points become evident.

To reduce unit water cost, that is cost in dollars per thousand gallons, it is essential to reduce the initial capital costs and the fixed charges. If cheap materials of construction are utilized, capital cost can be significantly reduced; but the operating and fixed charges may go up due to frequent material failures, the resulting down time, costly repairs and reduced plant life. Selection of the materials of construction has, therefore, to be based on a number of factors, the principle ones being the corrosion resistance and the initial cost.

For structural strength and easy workability, including repairs, most of the equipment in any plant is made from metals and alloys. Corrosion of those metals and alloys is undesirable since the product has no structural strength, is not heat conductive and finally the equipment may fail due to inadequate mechanical strength.

Corrosion is undesirable from several other viewpoints as well — corrosion products increase fluid flow resistance and hence aggravate pumping power requirements; corrosion products block flow passages and generate faulty process control signals so as to upset proper process regulations; accumulation of corrosion products on normally noncorroding metal surfaces initiates crevice or pitting type corrosion; liquid streams become polluted due to corrosion products; spillage and product loss and finally danger to safety of operating personnel exist when corrosion occurs. These are some of the reasons, both technical and economic, why corrosion must be avoided.

ENGINEERING STUDIES

In a two year study (2),(3) reported in 1968 entitled "Survey of Materials Behavior in Multi-Stage Flash Distillation Plants" sponsored by the Office of Saline

Water on materials behavior in desalination plants around the world, copper alloys showed the best performance record of all materials, a failure rate of much less than 1%.

Conducted by A.D. Little, this comprehensive study involved 55 plants from the Persian Gulf to California, a sample representing 75% of all existing desalting plants of the multistage flash distillation type, two years old or older and with a daily capacity of 100,000 gallons or more.

Emerson Newton, one of the authors of the study which included the service behavior of 70,000 heat exchanger tubes, recommended that Copper Alloy 715 (70-30 copper-nickel) be used for all stages of the evaporator and the brine heater if long term, trouble-free service was required. For maximum economy, this study recommended a tailored combination of three alloys: Copper Alloy 687 (aluminum brass) in the heat rejection stages, Copper Alloy 706 (90-10 copper-nickel) in the heat recovery stages, and Copper Alloy 715 (70-30 copper-nickel) in the brine heater.

Essentially all material failures experienced in the heat exchanger tubing were at the inlet ends of tubes and the result of either pitting or impingement where shells or other marine life lodged in the tube and caused excessive velocity at the point. However, tube failures were so rare that statistical analysis of the failures was not possible.

In 1966, two engineering feasibility studies (4),(5) to determine the cost of desalted water from dual-purpose nuclear power-water desalting plants were completed — one for a proposed seawater conversion plant of 150 million gallons per day capacity for the Metropolitan Water District of Southern California, and the other for a proposed 100 million gallon per day seawater conversion plant for the State of Israel.

In these studies, the unit cost of water was determined by dividing the annual cost allocated to water by the annual salable water production. Predictions of the annual salable water production required predicting desalting plant outage. Unlike predictions of electric power plant outages, which can be based on experience of other nuclear power plants, predictions of desalting plant outages for lack of experience with large desalting plants had to be based on an analysis of factors which may prevent operating the desalting plant at maximum capacity.

In preparing the analysis, assumption was made that the desalting plant had gone through all the procedures of "debugging" and all the failures due to faulty materials fabrication or assembly had been corrected.

The first component analyzed was the tubing in the heat reject and heat recovery stages and the brine heater. Tubing materials selected for these components were stainless steel-Copper Alloy 706, bimetallic in areas exposed to high concentrations of noncondensables and Copper Alloy 706 elsewhere. They were selected as reasonable compromises in first cost and expected performance.

Other materials considered were Copper Alloy 687 (aluminum brass) and 715 (70-30 copper-nickel). There is considerable information on a failure experience with Alloy 687 (aluminum brass) in power plant and marine condensers and this materials could give long service. However, the use of the more noble copper-nickel alloys would result in lower failure rates.

One of the best summaries on the subject is found in a report by Arthur D. Little on a survey of experience in 58 coastal power station condensers (6), which was carried out for the Office of Saline Water. The results shown in Table 1 are presented in terms of the probability of reaching 30 years of service in seawater in a power station condenser with less than 10% of lost tubes. The numbers range from a high of 85% probability of such outstanding performance for Copper Alloy 706 to a low of 30% probability for the admiralties, Alloys 443, 444 and 445. Current relative cost, based on $^{7}/_{8}$ inch 18 gauge tube are shown at the right.

TABLE 1: HEAT EXCHANGER SERVICE EXPERIENCE

Copper Alloy Number	Probability*	Relative Cost
706	85	1.28
715	81	1.53
687	57	1.00
443 - 445	30	0.68

* of reaching 30 years service in seawater
with less than 10% loss of tubes

On the basis of corrosion studies and experience, tubing throughout the seawater distillation plant could be expected to have an average corrosion rate of 0.001 inch per year on the seawater side and zero on the condensate side. An analysis of the tube failure rate based on the assumption that corrosion actually would be uniform at the average rate is completely unrealistic, since it assumes that the plant would be sound until the day of collapse. A statistical approach for predicting tube failures is much more realistic since there are on the order of a million tubes in the larger desalting plant designs. In order to perform this analysis, however, it is necessary to assume a tube failure rate and the forms by which failure will occur.

Failure Mechanisms

Even though a general thinning of tube walls should be expected over the life of a plant, some tubes will certainly develop leaks while the "average" tube is still sound. Mechanisms that can cause tube failure are as follows.

(1)   Rupture due to internal pressure of an abnormally thinned portion of a tube wall.

(2)   Pinhole leaks caused by localized pitting corrosion.

(3)   Breaks or cracks caused by vibration, improper metallurgical control or heat treatment, or by ammonia attack in the presence of tensile stresses.

(4)   Improper expansion of the tube into the tube sheet.

The third and fourth types of failures when they occur are likely to develop quite rapidly after a plant is placed in operation, but after the first year of continual operation, virtually all of these failures will have been discovered and corrective action taken.

Based on limited experience, it was assumed that 0.1% of the tubes in the heat reject and heat recovery stages and brine heater would fail during the first 10 years. This is consistent with the experience of the Los Angeles Department of Water and Power as one of their power plant condensers has operated for 53,000 hours and with none of the 800 Copper Alloy 706 (90-10 copper-nickel) tubes having failed, although some corrosion pits were observed. This observation gives better confidence in the 0.1% in 10 years failure rate assumption.

Since other values of tube failure rates may be assumed according to the judgement of the designer, the resulting plant operating factor, and therefore the unit cost of water are heavily influenced by this choice.

Probability of Failure

The probable number of failures each year is predicted according to the following distribution function:

$$F(t) = 1 - e^{-(t/A)^k}$$

where F is the probability of failure of a tube in t years, and k and A are constants; k (equals 2.5) being a constant which relates the distribution function to the corrosion phenomenon and A being a constant which relates the distribution function to the failure rate.

A calculation was made to determine the probable number of tube failues in any year, based on there being 940,000 tubes in the desalting plant, the assumption that 0.1% of the tubes would fail in 10 years of service, and that the salinity would not exceed 125 parts per million total dissolved solids at any time. A study of the availability of pumps and ejectors resulted in the inclusion of installed spares in the design and an availability of 1.0 could then be expected for these items. Structures could also be expected to experience no failures during the life of a plant. An availability of 0.99 was assumed for all other components. The combined availability and the resulting calculated desalting plant operating factor are shown in Table 2.

TABLE 2: PLANT AVAILABILITY

Years of Operation	Tubes	Ejectors, Pumps Structures	Auxiliary Equipment	Heat Source	Desalting Plant Operating Factor
1	0.992	1.0	0.99	0.90	0.884
2	0.986				0.879
5	0.973				0.867
10	0.959				0.856
15	0.951				0.847
20	0.942				0.839
30	0.911				0.812

Only normal operating conditions were considered in this analysis and no attempt was made to account for abnormal conditions such as major structural or equipment failures or pitting corrosion caused by small sea shells settling in the tubes, such as that shown in Figure 1, which failed due to erosion corrosion.

FIGURE 1: PITTING CORROSION OF HEAT EXCHANGE TUBE

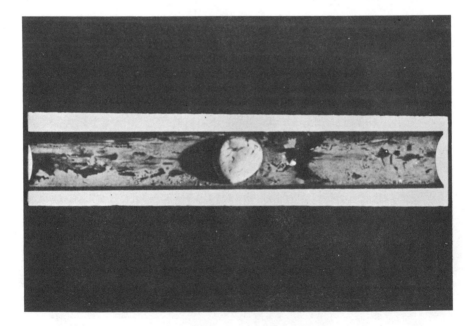

This was caused by a stone settling in the tube several inches from the inlet end.

COPPER DEVELOPMENT ASSOCIATION INVESTIGATION

Earlier, CDA reported the corrosion behavior of 12 copper alloys after up to 27 months exposure to seawater under conditions simulating the brine heater effect of a multieffect, multistage flash desalting plant (7)-(9). The investigation concurrently evaluated the influence of tube manufacturing method. By considering as-welded and scarfed seam-welded tube and also butt-welded seamless tube for possible future applications, certain potential cost savings could be experienced.

This investigation was carried out by The Dow Chemical Company under contract to CDA and with close cooperation with OSW.

This materials evaluation program initiated in December 1967 at the Materials Test Center in Freeport, Texas, was designed to simulate as much as possible the brine heater environment of the Clair Engle desalting plant. Because of the loop size, certain design restrictions were imposed; water flow to the loop could not exceed 9 gallons per minute, and no attempt was made to concentrate the water. It was found that, in order to match the Clair Engle surface area-to-volume ratio, addition of Copper Alloy /15 screening to the system was required. This was important since it concluded that the reduction in oxygen and the dissolved ions resulting from the corrosion of a given amount of surface area would influence the corrosivity of the water.

Tubular specimens fabricated from 12 copper and copper alloys together with their respective manufacturing methods and chemical compositions are shown in Tables 3 and 4.

TABLE 3: COPPER ALLOYS UNDER INVESTIGATION

Copper or Copper Alloy Number	Previous Tradename	Tube Manufacturing Method
122	Phosphorus Deoxidized, High Residual Phosphorus	Seamless
142	Phosphorus Deoxidized, Arsenical	Seamless
194	Copper-Iron Alloy	Seam Welded (SW)
443	Admiralty, Arsenical	Seamless
608	Aluminum Bronze	Seamless
613	Aluminum Bronze	Seamless
687	Aluminum Brass, Arsenical	Seamless
687	Aluminum Brass, Arsenical	Seamless, Butt Welded (BW)

(continued)

TABLE 3: (continued)

Copper or Copper Alloy Number	Previous Tradename	Tube Manufacturing Method
706	Copper-Nickel, 10%	Seamless
706	Copper-Nickel, 10%	Seam Welded (SW)
706	Copper-Nickel, 10%	Seamless Butt Welded (BW)
715	Copper-Nickel, 30%	Seamless
715	Copper-Nickel, 30%	Seam Welded (SW)
715	Copper-Nickel, 30%	Seamless Butt Welded (BW)
715*	Yorcoron	Seamless
716	Copper-Nickel 30% plus 5% Iron	Seamless
720	Copper-Nickel, 40%	Seamless

*Modified.

Test Conditions

In the loop, 20 inch long tube specimens were exposed to seawater at 250°F. and with a salinity of 80 to 100% of seawater but averaging 83%. The 20 inch specimen length was selected so that each specimen was long enough for inlet and outlet effects to be observed and so that enough tube surface was exposed to the hot seawater for localized corrosion. The specimens were still short enough to permit accurate weight loss measurements. Continuity of operation was interrupted only for short shutdowns for minor repairs and tube specimen removal.

Treated, deaerated seawater at pH 4 and a dissolved oxygen concentration of less than 5 parts per billion is supplied to the test loop from the nearby OSW's 100 gallon per minute treating plant. On entry to the loop, the water was adjusted to pH 7.4 with a 10% sodium hydroxide solution, and the dissolved oxygen is brought to the desired level by the addition of controlled amounts of oxygen-saturated seawater.

The dissolved oxygen in the feed seawater was approximately 72 parts per billion. This oxygen concentration was calculated in order that the CDA test loop would have the same mass flow of oxygen in the feed water per unit area of heat exchanger surface as would a typical brine heater section of an operating desalting plant with 20 parts per billion oxygen in the incoming feed water.

TABLE 4: CHEMICAL COMPOSITION OF ALLOYS

Composition, % Maximum (unless as a range or minimum)

Copper or Copper Alloy No.	Cu + Ag	Fe	Sn	Ni	Al	As	P	Pb	Mn	Zn	Total Other Elements
122	99.9	—	—	—	—	—	0.015-0.040	—	—	—	—
142	99.4	—	—	—	—	0.15-0.50	0.015-0.040	—	—	—	—
194	97.0	2.1-2.6	0.03	—	—	—	0.01-0.04	0.03	—	0.05-0.20	0.15
443	70.0-73.0	0.06	0.8-1.2	—	—	0.02-0.10	—	0.07	—	Rem.	0.15
608	99.5 (2)	0.10	—	—	5.0-6.5	0.02-0.35	—	0.10	—	—	—
613	99.5 (2)	3.5	0.20-0.50	—	6.0-8.0	—	—	—	—	—	—
687	76.0-79.0	0.06	—	—	1.8-2.5	0.02-0.10	—	0.07	—	Rem.	0.15
706	99.5 (2)	1.0-1.8	—	9.0-11.0	—	—	—	0.05	1.0	1.0	—
715	99.5 (2)	0.4-0.7	—	29.0-33.0	—	—	—	0.05	1.0	1.0	—
715 (1)	99.5 (2)	1.8-2.2	—	29.0-33.0	—	—	—	0.05	1.8-2.2	1.0	—
716	99.5 (2)	4.8-5.8	—	29.0-33.0	—	—	—	0.05	1.0	1.0	—
720	99.5 (2)	1.5-2.5	—	40.0-43.0	—	—	—	0.05	0.05-1.7	0.30	—

(1) Modified.
(2) Copper + elements with specific limits.

Schematic Layout of Loop

The schematic diagram of the CDA test loop shown in Figure 2 illustrates how the velocity effect was studied. Each of the 20 inch test bundles containing the 14 tube specimens was arranged in series and parallel paths so that the three water velocities of 16, 8 and 4 feet per second were obtained by dividing the total water flow to these bundles. They were not subject to heat exchange.

The two lower operating velocities (4 and 8 fps) of the test loop were selected as an average of a common industry design velocity of 6 fps. The 16 fps velocity is not encountered in practice but was selected as a means of evaluating the alloys under extreme velocity conditions.

FIGURE 2: SCHEMATIC LAYOUT OF CDA TEST LOOP

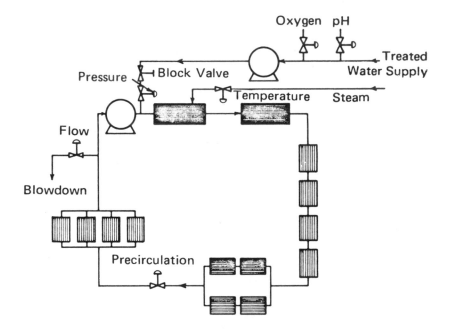

Velocity Bundles with Swagelok Fittings

Weight loss corrosion rates were determined from these bundles exclusively. Each tube in the test bundle was electrically insulated from the rest of the system by rubber sleeves between the tubes and the Swagelok fittings holding the tubes to the tube sheet as shown in Figure 3. These fittings allowed disassembly of the tube specimens from the tube sheets without damaging them, thus permitting accurate weight loss measurements.

FIGURE 3: VELOCITY TEST BUNDLE

Tube specimens 20 inches long by
¾ inch outside diameter are connected
to tube sheets by Swagelok fittings.

Leakage occurred through some of the Swagelok fittings causing some external surface corrosion of the tube specimens. Compensation for this external surface film weight was made by a procedure developed early in the program with this eventuality in mind.

Heat Exchange and Nonheat Exchange Bundles

Weight loss measurements were not possible on tubes from the heat exchanger and its companion nonheat exchange bundle because they were rolled into the tube sheets shown in Figure 4 making removal impossible without tube end damage. The tubes in the heat exchange bundle were 5 feet long. They were exposed to the hot seawater on the inside and 150 lb. steam on the outside.

FIGURE 4: HEAT EXCHANGER AND NONHEAT EXCHANGER

Tube specimens shown after 27 months exposure.

The nonheat exchange tubes were 20 inches long and had seawater on the inside but only stagnant air on the outside. Both bundles were jacketed with insulation for temperature control. Water velocity through these bundles was maintained at 5 feet per second.

The dissolved oxygen in the water leaving the heat exchanger was found to be below detectable levels, that is below 5 parts per billion. This level corresponds to actual operating plant brine heaters. It occurs as a result of the scavenging effect of heat exchange surfaces which the water contacts prior to reaching the brine heater.

Visually, surface films from the heat exchange and the nonheat exchange bundles were similar. Generally, they were very thin and tenacious with occasional bare metal showing through. Film color ranged the full spectrum, with combinations of colors appearing on the higher alloyed tubes.

Film Thickness

Film thicknesses for the alloys studied are shown in Table 5 for the 170 and 697 day exposure periods. At the 16 foot per second velocity, there is little difference in film thickness between the alloys. With the hot seawater velocity reduced to 8 feet per second, a slight increase in film thickness is seen particularly in the lower copper alloys. At the 4 foot per second velocity, a substantial increase in film thickness is observed on all alloys. The copper-nickel alloys in the 700 series, however, still possess the thinnest films of all the alloys.

TABLE 5: AVERAGE FILM THICKNESS (MILS) ON COPPER ALLOYS EXPOSED TO HOT SEAWATER FOR CONDITIONS INDICATED

Copper or Copper Alloy No.	Velocity—4 fps Time—Days		Velocity—8 fps Time—Days		Velocity—16 fps Time—Days	
	170	697	170	697	170	697
122	.35	(1)	.17	.24	.09	(1)
142	.38	.16	.12	(1)	.09	(1)
194 SW	.31	(1)	.12	.13	.06	(1)
443	.26	.08	.10	.19	.06	.05
608	.27	.32	.14	.44	.10	.24
613	.23	.30	.17	.13	.06	.13
687	.28	.08	.08	.08	.06	.05
706	.13	.08	.06	.07	.05	.05
706 SW	.12	.07	.06	.04	.05	.08
715	.11	.02	.06	.04	.04	.08
715 SW	.12	.02	.04	.04	.05	.02
715 mod	.13	.03	.07	.06	.06	.09
716	.17	.06	.06	.07	.05	.10
720	.12	.03	.04	.03	.04	(2)

[1] Film not measured.
[2] Questionable.

While the overall performance of the alloys in the two services was excellent, close examination suggests that tubes in the heat exchange bundle experienced slightly more attack than the corresponding nonheat exchange tubes in the form of general etching and micropitting. This micropitting was most prominent on tubes of Copper Alloys 443, 608, 613 and 687. These same alloys also displayed the least difference in performance between the two types of service. Where localized attack was noted in one service, it was generally apparent in the other. The micropitting, although fairly frequent, was generally quite shallow.

Figure 5 is a photomicrograph of a cross section through one of the deeper pits in the nonheat exchange Copper Alloy 613 tube. Its depth is 0.003 inch.

FIGURE 5: PHOTOMICROGRAPH OF A CROSS SECTION THROUGH THE DEEPEST PIT OBSERVED

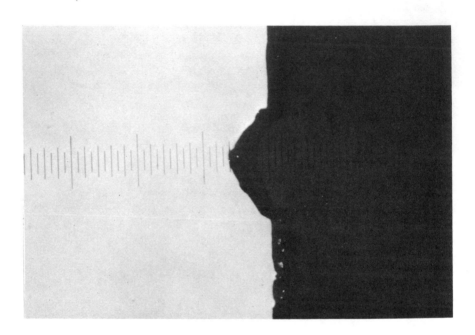

Depth 0.003 inch

Figure 6 is the inlet end tube sheet with rolled-in tubes of the five foot heat exchanger. The tube sheets are mild steel clad with Copper Alloy 706. No evidence of attack of the tubes or the copper alloy plate was noted.

Corrosion Rates

Weight loss data can be treated on either an "average" or an "instantaneous" basis. The instantaneous weight loss corrosion rates for the alloys tested are shown in Table 6.

At the 4 foot per second velocity, the data discloses that equilibrium had been achieved by the five alloys. At the higher velocities of 8 and 16 feet per second, these alloys are sorted out by order of merit under those conditions.

TABLE 6: INSTANTANEOUS CORROSION RATES FOR COPPER ALLOYS EXPOSED TO HOT SEAWATER

Corrosion Rate (mpy) for Conditions Indicated

Copper or Copper Alloy No.	Velocity—4 fps Exposure Time, Days				Velocity—8 fps Exposure Time, Days				Velocity—16 fps Exposure Time, Days			
	90	170	365	697	90	170	365	697	90	170	365	697
122	.55	.80	1.10	1.10	1.10	1.65	2.35	2.55	2.00	2.90	4.90	5.30
142	.80	1.20	1.45	1.50	1.30	.90	2.50	2.60	1.10	2.40	5.10	6.60
194 SW	.30	.60	1.05	1.10	1.00	.80	2.70	2.80	1.50	2.60	4.50	6.20
443	1.20	1.05	.60	.10	.70	.65	2.10	2.75	1.30	1.65	3.30	7.20
608	.50	.65	.80	.80	1.10	1.65	1.80	2.00	2.13	2.13	2.13	2.13
613	.50	.67	.67	.67	1.20	1.10	0.80	.30	1.70	2.80	4.40	5.10
687	.50	.50	.55	.55	.70	1.00	1.45	2.35	1.30	1.90	3.05	3.70
706	.70	.70	.71	.71	1.20	1.20	1.20	1.20	1.30	2.10	3.65	5.20
715	.40	.40	.40	.35	.20	.60	1.50	2.75	.85	1.60	3.20	4.45
715 mod	.80	.50	.20	.18	.50	.90	1.40	1.75	1.70	1.40	1.00	.80
716	.45	.45	.28	.28	.55	.90	1.40	1.70	1.30	1.85	2.20	3.90
720	.60	.55	.30	.28	.60	1.00	1.40	1.60	1.00	1.80	3.10	4.10

FIGURE 6: INLET END TUBE SHEET OF THE HEAT EXCHANGE BUNDLE

After 27 months exposure to 250°F. seawater

Although there was no evidence of erosion of the rolled-in tubes in the heat and nonheat exchange bundles at the 5 feet per second velocity, several of the tubes from the velocity bundles at the extreme velocity of 16 feet per second and which were electrically insulated did experience erosion.

The localized attack experienced at the inlet tube end of Copper Alloy 608 at the 16 foot per second velocity is shown in Figure 7. It is localized attack such as illustrated here which controls tube service life in practice.

Conclusions reached on this simulated brine heater test program are:

(1)   Most of the alloys tested will perform satisfactorily in this application.

(2)   Most of the alloys tested demonstrated lower corrosion rates with increased alloying content. Between the minimum and maximum average corrosion rates measured, little more than two mils metal loss per year represented the range of metal attack.

(3) Evidence of localized attack was seen on only some of the alloys and only at the extreme 16 feet per second velocity condition.

(4) To establish a theoretical service life, an assumption was made in that failure occurred when the tube wall thickness was reduced by corrosion to half of the original 0.049 inch. Corrosion rates calculated for the 23 month, 8 foot per second velocity were used to estimate service life. The minimum predicted service life for any of the alloys was with Copper Alloy 142 which was calculated at 20.1 years. Copper Alloy 720 had a predicted minimum life of 39.3 years with most of the alloys' anticipated service life exceeding 25.3 years.

(5) Seam-welded tube performance was excellent and showed negligible differences between it and their seamless counterparts. This was confirmed by the absence of selective weld and heat affected zone attack. These observations apply to both velocity and heat exchange bundles.

FIGURE 7: ERODED INLET END OF COPPER ALLOY 608 EXPOSED TO 260°F. 16 FPS SEAWATER

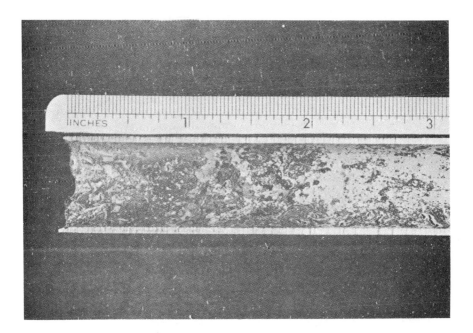

FREEPORT PLANT SECTIONS

The performance of the various materials under actual plant operating conditions
in the 1 million gallon per day OSW Freeport, Texas desalting plant over an eight
year period is summarized by plant section (10):

  (1)   Evaporator Effects

  (2)   Preheat Exchangers

  (3)   Final Condensers

  (4)   Piping

After approximately 18 months service, it was necessary to remove all the carbon steel evaporator tubes and to replace them with tubes of Copper Alloys
443, 687 and 706.

An inspection after a period of operation following the replacement indicated
that all of the replacement copper alloy tubes were in excellent condition. In
1967, inspection revealed that both Alloys 687 and 706 were performing well
but that Alloy 443 was deteriorating. The major cause of failure was attributed
to intergranular corrosion with evidence of impingement attack also present.
Some experimental fluted tubes of Copper Alloy 194 also showed no evidence
of corrosion.

Tube materials in the preheat exchangers were varied in the different exchangers
between carbon steel and Copper Alloys 443, 687 and 706. The relative general
superiority of the copper base alloys over steel was well demonstrated under the
conditions existing at the test locations. The performance of Alloy 443 was regarded as marginal to poor.

After five years of operation, no failures of the Alloy 715 cladding on tube
sheets and water boxes of nondeaerated seawater feed heat exchangers were
observed. One observer suggested that the Alloy 715 cladding was a contrib-
uting factor to the failure of the Alloy 443 tubes in one of the heat exchangers
since the failure occurred close to the tube sheet and the tubes were anodic to
the clad surfaces.

Trouble-free operation of the final condensers is very essential for maintaining
the final end temperatures since they control both plant capacity and plant production rate. This had both Alloy 687 and 706 tubes and no serious corrosion
problems were observed.

Piping in the Freeport plant transported six different fluids: seawater condensate,
vapor, brine blowdown, acid, alkali and raw seawater. It became very apparent
that repair and replacement of seawater and brine piping was the greatest single
maintenance item in the Freeport plant since all piping material was carbon
steel, only some of it being lined with a PVC or fiber glass epoxy coating. The
policy of using the least exotic metals in the plant was the primary emphasis

and was adopted as a design criteria with full knowledge that some materials would prove unsatisfactory and require early or frequent replacement. The process of evaluation thus established resulted in a final economic list of materials for future design considerations.

INTAKE AND DISTRIBUTION PIPING

While large diameter intake pipes to power plants are normally too large to be susceptible to clogging by massive marine growths, smaller pipelines must use chlorine to prevent this growth. Or in the case of one power company, their small diameter distribution pipeline had to be backwashed with hot water to eliminate fouling necessitating shutting down a portion of the system. Present day concern with the ecology is also a motivating factor for corrective action.

Figure 8 is a partially blocked pipe caused by corrosion product buildup and fouling. The use of either solid or clad copper or copper alloy pipe will contribute to the elimination of pipeline fouling.

FIGURE 8: PARTIALLY BLOCKED PIPE CAUSED BY CORROSION PRODUCT BUILDUP AND FOULING

TUBE SHEET AND WATER BOX

Figure 9 is a two-pass heat exchanger, the bottom half of the picture being the entry to the first pass, with much corrosion debris and tube pluggage being evident. But when a carbon steel water box is clad with a corrosion resistant material as seen in Figure 10, the absence of corrosion debris and tube plugging is noted.

FIGURE 9: ENTRY TO THE FIRST PASS OF A TWO-PASS HEAT EXCHANGER PLUGGED DUE TO CORROSION DEBRIS

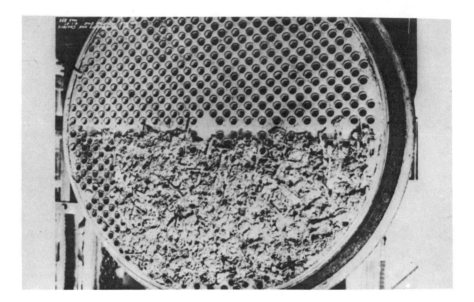

A successful spot welding procedure (11),(12) for attaching 0.025 inch thickness Alloy 706 to steel has recently been developed. This effort has received additional priority because of the realization that carbon steel piping water boxes and chamber walls were suffering far more severe corrosion in the deaerated seawater than had been previously anticipated.

This procedure makes it possible to both shop and field clad satisfactorily large areas of carbon steel with Alloy 706, which we believe will make their installed cost competitive with cement and organic linings which have yet to prove their long term durability in the desalting environment.

FIGURE 10: CARBON STEEL WATER BOX

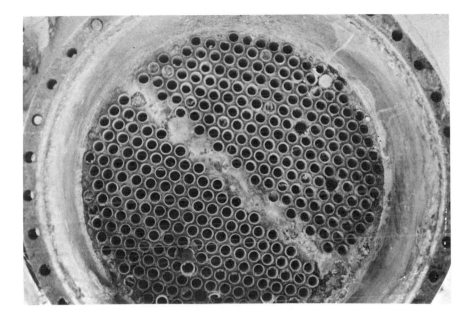

This has been clad with corrosion resistant material to prevent attack.

NEW CDA DESALTING PLANT

Early in 1971, Phase II of the CDA program was initiated to provide corrosion data on ten copper alloys in four additional desalting plant environments. The plant shown in Figure 11 went on stream in May 1971 and has operated continuously, producing over 5,000 gallons of fresh water per day with only one shutdown for tube specimen removal late in October.

Included in the plant are a heat recovery unit, flash evaporator, a vertical falling film exchanger, concentrated recycle brine unit and a side loop. Three test bundles from the initial test loop have been incorporated into the brine heater section of the plant in order to acquire a long term data point.

In the flow diagram shown in Figure 12 the treated seawater enters at the left, and flows successively through the heat recovery, brine heater, flash chamber, vertical tube exchanger, and finally the recycle brine section. The side loop using concentrated brine discharge is shown at the bottom of the figure.

In the heat recovery unit, the entering treated seawater is heated from 110° to 210°F. in five passes by the product water.

FIGURE 11: NEW COPPER DEVELOPMENT ASSOCIATION DESALTING PLANT

Capacity approximately 6,000 gallons per day

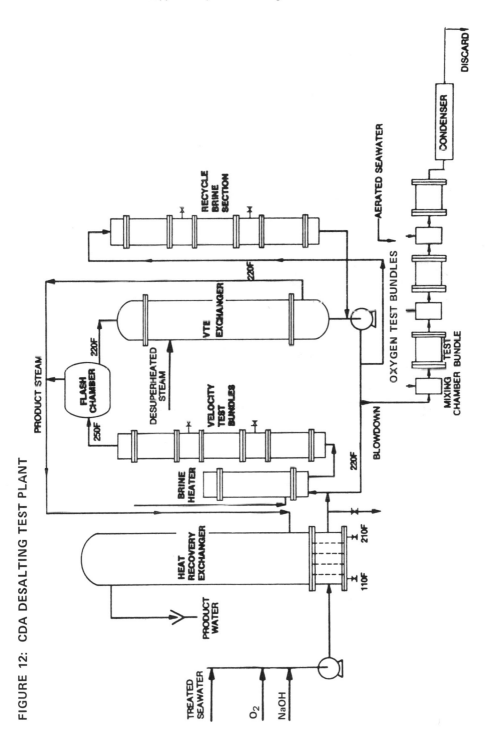

FIGURE 12: CDA DESALTING TEST PLANT

Oxygen concentration and temperature are periodically measured at a number of locations within the plant with the Petrolite unit shown in Figure 13, to enable corrosion rates to be related directly to these variables for the ten copper alloys under study.

FIGURE 13: PETROLITE UNIT

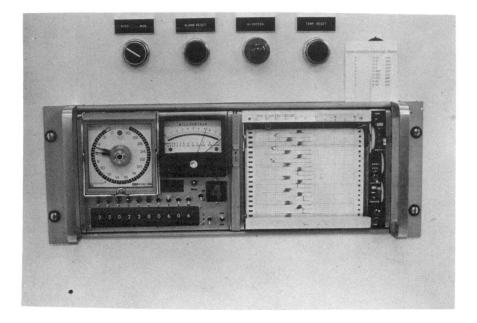

This monitors and controls oxygen concentrations to preset conditions.

In addition, a side unit shown in Figure 14 is using the concentrated brine discharged from the test plant to provide short term corrosion data on the effects of oxygen content at known temperatures on a group of selected alloys without disrupting the operating conditions in the main plant. This unit consists of three 4 tube test bundles in series with provisions for oxygen injection, mixing, monitoring and recording at the entrace to each exchanger.

The data when finalized will be presented at the Desalination Symposium of the annual meeting of the National Association of Corrosion Engineers.

FIGURE 14: SIDE UNIT

Used for short term testing of copper alloys under extreme environmental conditions.

REFERENCES

(1)   F.L. LaQue and H.R. Copson, Corrosion Resistance of Metals and Alloys, Second Edition, Reinhold Publishing Corporation, New York, 1963, pp. 575-599.

(2)   E.H. Newton and J.D. Birkett, "Survey of Materials Behavior in Multi-Stage Flash Distillation Plants. " Report of the Office of Saline Water, Department of the Interior, Washington, D.C., August 1968.

(3)   E.H. Newton and J.D. Birkett, "Summary Report on Survey of Materials Behavior in Multi-Stage Flash Distillation Plants. " Symposium sponsored by the Office of Saline Water, Department of the Interior, Washington, D.C., September 25, 1968.

(4)   N. Arad, S.F. Mulford and J.R. Wilson, "Desalting Plant Down Time Predicted by Formula. "Environmental Science and Technology, Vol. 2, No. 6 (June, 1968), pp. 420-427.

(5)   N. Arad, S.F. Mulford and J.R. Wilson, "Prediction of Large Desalting
        Plant Availability Factor," Second European Symposium on Fresh
        Water from the Sea, May, 1967.

(6)   E.H. Newton and J.D. Birkett, "Survey of Condenser Tube Life in
        Salt Water Service, " Report No. C-68373 Office of Saline Water,
        Department of the Interior, February, 1967.

(7)   A. Cohen and L. Rice, Materials Protection, Vol. 8 (December, 1969),
        pp. 67-69, and Proc. NACE 25th Conference 342-345, NACE, Houston,
        Texas.

(8)   A. Cohen and L. Rice, Materials Protection and Performance, Vol. 9
        (November, 1970), pp. 29-35.

(9)   A. Cohen and A.L. Whitted, Materials Protection and Performance,
        Vol. 10 (November, 1971), pp. 34-37.

(10)   "Final Report of An Analysis and Summary of Reports and Data
        Freeport Test Bed Plant for the Period 1961-1969." Office of Saline
        Water, U.S. Department of the Interior, Washington, D.C., September
        1971.

(11)   W.F. Ridgeway and D.J. Heath, "Lining Mild Steel Components with
        90:10 Copper-Nickel Alloy Sheet." Welding and Metal Fabrication
        Journal, October 1969.

(12)   Private Communication — The International Nickel Company.

DESALINATION — LOWER COST WATER BY PROPER MATERIALS SELECTION

B. Todd, International Nickel Limited
A.H. Tuthill, The International Nickel Company, Inc.
R.E. Bailie, Desalting Systems and Services, Inc., Fort Lauderdale, Florida

Desalination Conference, Yugoslavia
September 14-17, 1970

ABSTRACT

This paper considers the reasons for desalination plant shutdown and shows how proper selection of materials can minimize downtime and lead to lower water costs. Particular attention is paid to the selection of materials for key components in the water intake system and in the head recovery section. Materials for pumps and valves are also discussed.

GENERAL

Lower cost water can be achieved by increasing the percentage of time a desalination plant actually operates and produces salable water. Numerous studies and operating experience have shown that increasing the percentage of on-stream time effects a greater reduction in the cost of water than any other major variable under the control of the plant designer and owner's engineer.

Newton and Birkett in Table 1 list the reasons for desalination plant shutdown in the approximate order of their importance. General maintenance, intake structure maintenance and water box maintenance each caused more shutdowns than tube replacement.

TABLE 1: REASONS FOR PLANT SHUTDOWN OR DECREASED PRODUCTION IN APPROXIMATE ORDER OF IMPORTANCE

(1) Scale removal — acid cleaning
(2) General maintenance
(3) Intake structure maintenance
(4) Water box repair

(continued)

71

TABLE 1: (continued)

    (5) Tube plugging and replacement
    (6) Temperature imbalance
    (7) Imbalance in steam distribution between power plant and evaporator — for dual purpose plants
    (8) Steam boiler problems — for single purpose plants
    (9) Adjustments of venting, vacuum, pressure lines
    (10) High salinity product
    (11) Pump troubles
    (12) Brine heater malfunction — repairs
    (13) Feed treatment — equipment problems
    (14) Low water demand
    (15) Brine flow — recirculation problems
    (16) Nozzle and flashing problems

Table 2 reveals the cost of downtime to be about $5,000 per day for a 2.5 mgd plant in terms of the value of the product water alone. The true cost of the water lost is but a small part of the real cost if the community depends on the desalination plant for a significant portion of its water supply.

TABLE 2: COST OF DOWNTIME IN DESALINATION PLANTS

Basis — 2.5 mgd plant selling water for $2.00
    per 1,000 gals.
Loss of revenue — $5,000 per day of downtime
Maintenance and operating labor at $5,000
    month — $167.67/day
Maintenance and operating material at $2,000
    month — $67.67/day

In many arid areas of the world lack of skilled maintenance already represents an almost insurmountable barrier to desalination as a source of fresh water. Considerable attention must be given to increasing the durability of a number of components of desalination plants, besides the turbine, if desalination is to meet the growing needs of mankind for new sources of water.

Examination of Table 1 reveals that apart from scaling (which is primarily associated with the brine heater) and operational problems, the repairs and maintenance causing shutdown occur principally in the inlet, or heat reject, section of the plant.

The tubing, pipe, water boxes and intake structure of the inlet section also suffer the greatest abuse from exposure to the incoming, raw seawater. However, the brine piping, pumps, water boxes and even the chamber walls of the heat recovery section also suffer abuse despite the fact that they are exposed only to treated, deaerated seawater brine. Heat recovery section equipment failures also contribute substantially to shutdowns.

Operating records throughout the world have clearly confirmed that improper selection of materials for components of the piping systems in both the inlet and heat recovery sections of the plant can transform an otherwise reliable desalting facility into an operational failure with ever-increasing downtime and loss of production, not to mention the psychological effect on potential owners who are evaluating desalting processes for their water supply.

Heat Reject Section

This section serves to preheat the raw seawater while also rejecting low level heat from the system. Depending on the process design, the seawater leaves the heat reject section at from 95° to 115°F. (35° to 46°C.). When warmed to these temperatures and treated with acid, the dissolved gases are readily removed.

Deaeration is accomplished either in a combination vacuum decarbonator/deaerator or in a two-stage degasification system utilizing an atmospheric decarbonator, then a vacuum deaerator.

Plant Location

One early decision the owner's engineer must make overshadows all others in determining the downtime and maintenance that the inlet section of the plant and the plant itself will be subject to. This is the specific location chosen for the intake. The amount of marine life such as mussels, seaweed, kelp and small fish, and the amount of debris such as seashells, crab claws, sand, leaves and even stones that enter the intake with the seawater vary widely from place to place.

Normal seasonal variations in the outflow of estuaries, tides and seasonal variations in the sea itself result in fluctuations in the type and quantity of this extraneous matter in the seawater which is to be desalted. The engineer has three basic options. He can take seawater from:

      (1)   the open sea, bay or estuary by running a pipe out from shore with a vertical riser and intake screen;

      (2)   an existing back bay, man-made channel or natural tidal estuary;

      (3)   a beach well using the beach sands as a natural filtration plant.

If the intake is from the open sea or from a channel, a screen must be provided to remove debris, and heavy chlorination must be used to control marine growth.

Some sand, shell fragments, crab claw parts, etc., will pass through even the best screening systems and be carried to the tubing; therefore, allowance must be made for periodic cleaning to remove them and the marine growth that chlorine injection may not entirely prevent.

The seawater supply piping to and from the heat reject tube bundle should be arranged with provisions for off-stream chemical cleaning. This can be accomplished easily and economically at the time plans and specifications are being prepared, but can present numerous problems if modifications have to be made after the plant has been commissioned.

If beach wells are chosen, the beach sands will act as a filter and provide a clean, well-filtered seawater feed. Some sand and/or coral is normally encountered but the debris and marine life are effectively removed. Unfortunately, a new but very real problem arises — ammonia and sulfides from decay of organic life and organic wastes. In passing through the beach sands which are all too often rich in decomposed organic matter, the dissolved oxygen can be stripped from the seawater and replaced with hydrogen sulfide.

It is possible to properly space and locate beach wells so that relatively little reduction of the normal dissolved oxygen content of seawater occurs as the seawater is drawn through the beach sands. Unfortunately this has not always been achieved in practice. In some island locations rainfall has brought deaerated, decay-enriched groundwater intrusion into beach wells from the land slide.

In other locations, it has been found that the decay products (sulfides and ammonia) persist even after several years of continuous pumping. Properly locating beach wells requires a high degree of skill and detailed hydrological studies.

There are methods of treating a persistent hydrogen sulfide (H_2S) problem. Conventional water treatment techniques can be applied. Any treatment, however, results in a more costly seawater system. Upgrading of the tube and tubesheet material is an alternative to the cost of treatment to remove hydrogen sulfide, but also results in a more costly heat reject section. Care in making the detailed hydrological studies necessary to insure continuous flow of aerated seawater to each beach well will avoid the cost of treatment and the use of more expensive tubing alloys.

Whatever debris, pollutants or fish are brought by the winds, tides and currents to the intake structure will either be screened out or be passed through the desalination plant. Table 3 shows two water analyses. The analysis at the top is the conventional one the analyst provides when the plant designer asks for a water analysis. The analysis at the bottom, which is termed the Stick and Stone Index, is really a list of the root causes of much of the shutdown and maintenance in the inlet section.

Table 4, a survey of reported tube failures, indicates how debris and dissolved gases contribute to condenser tube failures.

TABLE 3

Water Analysis

Ion	Parts per Million
Chloride	18,980.0
Sulfate	2,649.0
Bicarbonate	139.7
Bromine	64.6
Fluoride	1.3
Boric acid	26.0
Sodium	10,556.1
Magnesium	1,272.0
Calcium	400.1
Potassium	380.0
Strontium	13.3

Stick and Stone Index (SSI)

Passing Screen

Sticks, stones, crab claws, seashells, sand, gravel, fish, plankton, sewage and leaves

Process Additions

Chlorine, acid, caustic, polyphosphate

Contributions from Internal Pipe and Water Box Walls

Rust flakes, spalled linings, mussel shells

Dissolved Gases and Ammonia

Oxygen, carbon dioxide, hydrogen sulfide and ammonium ions

From Decay of Plankton in Recycling Brine

Hydrogen sulfide and ammonium ions

TABLE 4: A SURVEY OF REPORTED TUBE FAILURES

Condensers are normally retubed when the number of plugged tubes approaches 10%. When some 2 to 3% of the tubes are plugged, the engineer usually has enough lead time to procure new tubes for orderly replacement. The greater portion of all failures are in the vicinity of the inlet end, with down the tube failures and external failures responsible for the balance.

CAUSES OF INTERNAL OR WATER SIDE FAILURES

	90-10 Cu-Ni-1.5 Fe Alloy	70-30 Cu-Ni-0.5 Fe Alloy
DESIGN CONDITIONS — UNIFORM GENERAL CORROSION	0.1 to 0.5 mpy (0.0025-0.0125 mm./yr.)	0.1 to 0.5 mpy (0.0025-0.0125 mm./yr.)
This refers to the very slow barely measurable general corrosion of copper base condenser tube alloys. It is generally assumed that the rates are controlled by diffusion processes through protective film.		
DEVIATIONS FROM DESIGN CONDITIONS — ACCELERATED GENERAL CORROSION		
Pollution	Less resistant	Preferred for polluted waters
Sewage (O_2 depletion, H_2S increase, NH_3 possibly present).		
Marine organisms, for example: Red Tide; Gulf of Kuwait. (O_2 depletion, H_2S increase, NH_3 possibly present).		
Industrial wastes (acidic wastes are chief problem).		
Selective attack	Naturally resistant	Naturally resistant but selective attack reportedly can occur at very high temperatures. See hot spot attack.
Dezincification of brass tubes was an early cause of failure which has been virtually eliminated except in certain refinery condensers operating at high temperatures. The inhibitors developed for the purpose and added to the tube metal have been quite effective at normal condenser operating temperatures.		
Water side velocity above limit (Common cause of failure)	8 to 12 fps (2.5-3.8 mps) Iron below 1.0% lowers tolerable velocity markedly	15 to 20 fps (4.5-6 mps) Iron below 0.4% lowers tolerable velocity markedly
As seawater velocity increases, the protective film is stripped away from copper base condenser tube alloys. Experience has shown the limiting design velocity varies for each alloy and is a key factor in condenser tube selection. The design velocity should be somewhat below the limiting velocity established in test work to allow for eddies and turbulence.		

(continued)

TABLE 4: (continued)

		90-10 Cu-Ni-1.5 Fe Alloy	70-30 Cu-Ni-0.5 Fe Alloy
Cleaning of deposited scale (Minor cause of failure)	Both chemical cleaning and mechanical scraping methods have been used to remove heavy deposits of scale that reduce heat transfer in certain condensers operating in unfavorable waters. Poor control has led to tube failure.	Survives most cleaning methods	Survives most cleaning methods
Abrasion from entrained sand (A problem in certain locations)	The protective film is apparently scoured away by the abrasive action of the sand particles at velocities that are otherwise tolerable for the particular tube alloy.		Use 2% Fe alloy

MORE LOCALIZED EFFECTS

		90-10 Cu-Ni-1.5 Fe Alloy	70-30 Cu-Ni-0.5 Fe Alloy
Inlet and erosion (Major cause of failures)	Pollution, velocity effects and abrasion tend to affect the inlet end first. Water box designs that permit impingement and turbulence effects at the tube ends or undesirable galvanic effects are particularly troublesome.	8 to 12 fps (2.5-3.8 mps) Iron below 1.0% lowers tolerable velocity markedly	15 to 20 fps (4.5-6 mm./sec.) Iron below 0.4% lowers tolerable velocity markedly
Galvanic effects	The use of steel waster plates, zinc or magnesium anodes or bare steel or bare cast iron water boxes provides a substantial measure of cathodic protection to the tube ends, which is particularly helpful to brass alloys with lower tolerable velocity limits.	Water box design is critical. Waster plates (steel or zinc) are not required if the copper-nickel tubes are of the correct iron content and if the water box itself is designed to avoid excessive turbulence. Unsatisfactory water box designs and lower Fe content tubing can often be offset by properly installed zinc or steel waster plates. Care must be taken when waster plates are used that these do not themselves introduce excessive turbulence.	
Nonmetallic water box and pump linings	Spalling and peeling of pump and water box liners lead to partial or complete plugging and/or blocking of individual tubes, deposit attack and hot spot corrosion.	Not resistant	Not resistant
Seashells and crab claws	Broken seashells and crab claws entering the tubes often lodge therein, partially plugging the tube.	Good resistance. Some failure reported	Better resistance. Few failures reported

(continued)

TABLE 4: (continued)

	90-10 Cu-Ni-1.5 Fe Alloy	70-30 Cu-Ni-0.5 Fe Alloy
Deposit attack Leaves, seaweed, corrosion scale and other entrained matter tend to deposit in the bottom of the tube, particularly during any period when the actual flow rate in individual tubes drops to 3 fps (1 mps) or less.	Insufficient data to rate	Resistant. A few failures are reported in refinery condensers with condensing temperatures above 400°F. (200°C.)

STRESS CORROSION CRACKING

	90-10 Cu-Ni-1.5 Fe Alloy	70-30 Cu-Ni-0.5 Fe Alloy
Traces of ammonia, moisture and air have led to stress corrosion cracking of brass alloy condenser tubes even during construction and startup. In operation cracking has occurred in the air ejector section of power plant condensers and in coolers operated in certain river waters in the U.S. and Japan.	Resistant	Resistant

CAUSES OF EXTERNAL OR STEAM SIDE FAILURES

	90-10 Cu-Ni-1.5 Fe Alloy	70-3- Cu-Ni-0.5 Fe Alloy
ACCELERATED GENERAL CORROSION Exfoliation The external surfaces of 70-30 copper-nickel tubing and 80-20 copper-nickel tubing in high temperature (200°C. plus) feedwater heaters retired to "peaking" service have suffered general exfoliation-type attack when the units are repeatedly cycled between operating and ambient temperatures and when air (O$_2$) is admitted to the steam chamber.	Resistant	Susceptible at high temperatures
MORE LOCALIZED EFFECTS Steam impingement Wet steam entering the condenser at substantial velocities may "cut" the outer row of tubes unless baffled.	Primarily a design factor though resistance to steam cutting varies somewhat	

(continued)

TABLE 4: (continued)

	90-10 Cu-Ni-1.5 Fe Alloy	70-30 Cu-Ni-0.5 Fe Alloy
Vibration	Primarily a design factor	
Tube failures occur from vibration near the tube sheets and sometimes at the support plates. The vibrations arise from the velocity of the entering vapors and the spacing of the intermediate support plates between tube sheets.		
Noncondensable gases	Resistant	More resistant
Ammonia and carbon dioxide in the condenser have led to both cracking and localized corrosion of the external surfaces of brass alloy tubes in power plant condensers.		
	Less resistant	Resistant
H_2S when present has led to corrosion of external surfaces of tubes in deaerator and certain sections of heat recovery section of desalination plants. Sulfides in hydrocarbon streams have led to external corrosion of tubes at high temperatures.		

Heat Recovery Sections

The treated and deaerated seawater from the heat reject section joins the recycle brine stream and enters the heat recovery section (the longest section in the plant) at 95° to 115°F. (35° to 46°C.). As it flows through the multiple stages of this section it is warmed by the product water vapors condensing on the outside of the tubes in each stage. The exit temperature reaches 250°F. (99°C.) in modern acid-treated plants.

Heavy maintenance on carbon steel piping and water boxes of the lower temperature (below 160° to 180°F.) stages of the heat recovery section can be deferred to the later years of operation by careful control of the pH, residual dissolved oxygen, carbon dioxide and hydrogen sulfide levels in the brine leaving the deaerator; care in avoiding the lower pH's during on-stream acid cleaning; and a fully trained operating staff.

Disregard of any of the above factors will lead to shutdowns and heavy maintenance at an earlier time.

Cost of Downtime in Desalination Plants

Basis — 2.5 mgd plant selling water for $2.00 per 1,000 gal.

Loss of revenue — $5,000 per day of downtime

Maintenance and operating labor at $5,000 month = $167.67/day

Maintenance and operating material at $2,000 month = $67.67/day

It is clear that the overriding item is the loss of revenue and not the maintenance and operating labor and material costs which are, in any case, incurred. Thus, keeping the plant operational two extra days per month provides enough additional revenue to pay for the regular maintenance and operating labor force.

90-10 copper-nickel alloy pipe and water boxes are already in common use in the higher temperature stages above 180°F. (80°C.). Details of the materials used are given in Table 5.

TABLE 5: COMPOSITIONS OF MATERIALS MENTIONED IN THE PAPER

Stainless Steels

Steel	Nominal Composition Percent					
	Carbon (max.)	Chromium	Nickel	Molybdenum	Copper	Others
AISI Type 304	0.08	18.5	9.5	–	–	–
AISI Type 304L	0.03	18.5	10	–	–	–
AISI Type 316	0.08	17	12	2.5	–	–
AISI Type 316L	0.03	17	13	2.5	–	–

(continued)

TABLE 5: (continued)

Steel	Carbon (max.)	Chromium	Nickel	Molybdenum	Copper	Others
			Nominal Composition Percent			
ACI-CF3	0.03	19	10	–	–	–
ACI-CF3M	0.03	19	11	2.5	–	–
ACI-CF8	0.08	19	9	–	–	–
ACI-CF8M	0.08	19	10	2.5	–	–
ACI-CD4M Cu	0.04	26	5.5	2.5	3	–
ACI-CN7M (Alloy 20)	0.07	20	29	2	3	–
Carpenter 20 Cb-3*	0.07	29	20	2.5	3.5	0.6 Cb
Worthite**	0.07	20	23	3	–	3.5 Si

Ni-Resist Cast Irons

Alloy	C (max.)	Si	Mn	P (max.)	Ni	Cu	Cr
					Composition Percent		
Type I	3.0	1.0-2.8	1.0-1.5	–	13.5-17.5	5.5-7.5	1.75-2.5
Type IB	3.0	1.0-2.8	1.0-1.5	–	13.5-17.5	5.5-7.5	2.75-3.5
Type II	3.0	1.0-2.8	0.8-1.5	–	18.0-22.0	0.5 max.	1.75-2.5
Type D-2	3.0	1.75-3.0	0.7-1.0	0.08	18.0-22.0	–	1.75-2.5
Type D-2B	3.0	1.75-3.0	0.7-1.0	0.08	18.0-22.0	–	2.75-4.0

Alloys of Nickel and Copper

Alloy	Copper	Nickel	Iron	Other
		Nominal Composition Percent		
90-10 Cu-Ni	Remainder	10	1.5	1.0 Mn (max.)
70-30 Cu-Ni	Remainder	30	0.6	1.0 Mn (max.)
70-30 Cu-Ni (High Iron)	Remainder	30	2.0	2.0 Mn
Ni-Cu Alloy 400	31.5	66	1.35	0.9 Mn
Ni-Cu Alloy 410	30.5	66	1.35	1.6 Si
Ni-Cu Alloy K500	31.5	66	1.35	0.9 Mn, 2.8 Al, 0.5 Ti
Ni-Cu Alloy 506	30	64	1.5	3.2 Si

Nickel-Chromium-Iron Alloys

Alloy	Carbon	Chromium	Nickel	Iron	Molybdenum	Other
			Nominal Composition Percent			
Inconel*** X-750	0.4	15	Remainder	6.75	–	2.5 Ti, 0.8 Al, 0.85 Cb
Inconel Alloy 625	0.05	18.5	Remainder	3	9	4 Cb
Incoloy*** Alloy 825	0.03	21.5	41.5	30	3	1.8 Cu, 0.9 Ti, 0.15 Al

*Trademark, Carpenter Technology Corporation
**Trademark, Worthington Corporation
***Trademark, The International Nickel Company, Inc.

Demisters serve to strip droplets of liquid from the vapors as they pass from the flashing brine in the bottom of flash chamber to the tubing in the top where the vapors condense. Brine droplets that carry through the demisters reduce product water purity.

Monel Alloy 400 demisters are the most widely used, but in some plants with high hydrogen sulfide content in the feedwater, demisters in this alloy in the first few higher temperature stages as well as the last stages of the heat reject section have suffered enough corrosion to require replacement.

Both Type 316 stainless steel and Incoloy Alloy 825 have been used to replace Alloy 400 in these stages, with apparent success to date.

MATERIALS SELECTION

The following sections indicate materials believed to achieve maximum improvement in on-stream time over the full 20 to 30 year lifetime of modern 1 to 10 million gallon per day desalination plants, at minimum increase in cost.

Intake Pipe to Open Sea

The need for periodic removal of marine fouling from the interior suggests a minimum diameter of 6 to 8 feet (1.8 to 2.4 meters) and a requirement for a diver manhole every 50 to 100 feet (15 to 30 meters). If seawater penetrates the concrete and reaches the steel reinforcing bar, the pressure from the volume of the corrosion products can cause the concrete to spall.

Care must be taken, in selecting the concrete formulation and in pouring, to achieve a dense seawater resistant structure. Care must also be taken to insure that all portions of the reinforcing bar are fully covered with 2 inches (50 mm.) or more of concrete.

Some of the copper family of materials, such as silicon bronze, copper or brass, are incompatible galvanically with the more noble 90-10 copper-nickel of the vertical riser and should be avoided. Nickel-copper Alloy 400 should be used (see Table 6).

TABLE 6: INTAKE PIPE TO OPEN SEA

Horizontal run	Concrete
Vertical Riser	CA706 (90-10 copper-nickel)
Fasteners and miscellaneous hardware	Nickel-copper Alloy 400

Trashracks, Stationary Screens, Traveling Screens and Rotary Screens

Trashracks, stationary screens, traveling screens and, more recently, microstainers are used to reduce fouling and plugging of the tubes and to keep heat transfer rates up to new "clean-tube" valves (see Table 7). Trashracks are commonly steel, though there has been some experimentation with light gauge alloy designs. The screens and the frames holding the screens are commonly Type 304 or 316 stainless, but the structural members are usually coated carbon steel.

TABLE 7: MATERIALS FOR COMPONENTS OF TRAVELING WATER SCREEN FOR SEAWATER

Component	Premature Failures Reported[1]	Minimum Materials for Moderate Maintenance[2]	Upgraded Materials for Lowest Maintenance[3]
Main head terminal cover	Carbon steel, coated	Type 304 (CP)[4] fiberglass	CA 706 (90-10 Cu-Ni)[5]
Rivets, bolts and staybolts for main supporting frame	Type 303 (18/8 Cr/Ni + S) Type 304	Type 316 (CP)	Nickel-copper alloy 400
Screen panel frame	Carbon steel, coated	Type 304 (CP)	CA 706
Screen wire retaining frames		Type 304 (CP) CA 706	Nickel-copper alloy 400
Screen wire[6]		Type 316 (CP)	Nickel-copper alloy 400 or K500
Head and foot shafts		Type 304 (CP)	Nickel-copper alloy 400 or K500
Chain sprocket wheels	CF8 (316)	Ni-Resist CN7M (Alloy 20) Nickel-aluminum bronze	
Shaft bushings and foot sprockets		Leaded tin bronze	
Sleeve shaft bearings		Leaded tin bronze	
Kelp knives		Type 316	
Lifting lips		Type 316	

(continued)

TABLE 7: (continued)

Component	Premature Failures Reported[1]	Minimum Materials for Moderate Maintenance[2]	Upgraded Materials for Lowest Maintenance[3]
Bottom boot		Type 304 (CP)	CA 715 (70-30 Cu-Ni)
Splash plates		Type 304 (CP)	CA 706
Main chain rollers	Type 440-B	Type 304 (CP) Ppt Hardening S.S.	Nickel-copper alloy K500
Main chain side bars	Type 410	Type 304 (CP)	Nickel-copper alloy K500
Main chain roller guides	Type 420 (0.15% C min.)	Type 304 (CP) Ppt Hardening S.S.	Nickel-copper alloy K500
Main chain bushings	Type 420	Type 304 (CP) Ppt Hardening S.S.	Nickel-copper alloy K500
Main chain pins	Type 420	Type 304 (CP) Ppt Hardening S.S.	Nickel-copper alloy K500
Attachment bolts in baskets		Type 316 (CP)	Nickel-copper alloy K500
Cap screws for baskets		Type 316 (CP)	Nickel-copper alloy 400
Take-up screws	Type 416	Type 316 (CP)	Nickel-copper alloy 400
Take-up screw ball bearing		Oil sealed bearing	
Take-up screw nuts		Type 316 (CP)	Nickel-copper alloy 400
Spray nozzles		Type 304 (CP)	Nickel-aluminum bronze
Spray header		Type 304 (CP)	CA 706

(continued)

TABLE 7: (continued)

Component	Premature Failures Reported[1]	Minimum Materials for Moderate Maintenance[2]	Upgraded Materials for Lowest Maintenance[3]
Screen guideways		Type 304 (CP) or Concrete	Ni-Resist CA 715
Driving mechanism housing		Type 304 (CP)	CA 706
Extended track shoes	CA 15 (410)	Cast nickel-copper alloy 505	
Anchor bolts		Nickel-copper alloy 400	
Lubrication fittings		Type 316 (CP) Nickel-copper alloy 400	

[1]Reports indicate corrosion of components in these materials has led to early replacement and substantial maintenance.

[2]These materials appear to be the minimum upgrading to keep maintenance within reasonable limits in coastal plants with good maintenance facilities.

[3]Some installations have upgraded the components indicated to effect further reduction in maintenance. More isolated desalination plants should consider such further upgrading. Where no material is listed in this column, it means that further upgrading appears unnecessary.

[4]CP following an item means that cathodic protection is required to overcome crevice corrosion and pitting.

[5]CA is the Copper Development Association designation for the cast or wrought copper alloy (CAxxx).

[6]Copper alloy screens suffer corrosion particularly where automatic high pressure backwashing is fitted. Life of copper alloy screens has been reported as directly related to number of washings. Copper alloy screen frames on the other hand are particularly appropriate as they provide cathodic protection to the more noble materials used for the wire.

The stainless steel screens and frames usually receive sufficient galvanic protection from the structural steel, or from the sacrificial anodes often employed, to over- come their tendency toward crevice corrosion. Unlike the thin wall piping and water boxes (which can be reinforced externally), light-weight alloy frame de- signs have not yet been evolved for this equipment.

The owner must rely on coatings and cathodic protection not only for the frames but for the stainless steel screens, chains and other working components as well. The screens, sprockets, bearings, chain and other components in the commonly available stainless alloys will give good performance when supplied with ample cathodic protection. Crevice corrosion and high maintenance can be expected in the absence of ample cathodic protection.

Collecting Basin or Tank

Downstream of the traveling screens there is usually a basin or surge tank made of concrete or coated steel, which serves as a reservoir of screened or filtered water. The basin should be sized to permit as much as possible of the detritus that passes through the screens to settle out.

Arrangements can be made to spray and aerate the incoming water to the col- lecting basin or holding tank to vent and reduce the H_2S content. Nonmetallic spray nozzles are suggested. The outflow to the pump can be taken from the top behind a baffle to encourage further settling. Provision should be made for periodic removal of debris that settles to the bottom.

There has been experimentation with 90-10 copper-nickel linings of concrete holding basins to reduce marine fouling on the side walls and bottom, but re- sults to date are inconclusive.

Piping, Water Box and Tubing

Coatings look good and perform well initially when properly applied. After several years they all too often begin to fail. As failure begins, the coating spalls off in small or large sections. If the sections of coating are not large enough to block the pump intake, they pass through the pump, lodge in the tubes and block them, causing tube failures and downtime.

Even cathodic protection of internal surfaces may introduce more problems than it solves. The impressed current systems require special controls and skilled at- tention to avoid overprotection, which blisters coatings. Polarity may be re- versed, and occasionally has been, after routine maintenance. In some instances the potential has drifted high enough to actually decompose seawater and intro- duce substantial chlorine contents.

Sacrificial anodes will not operate in reverse nor will they generate chlorine. Being bulky, they may restrict, and occasionally have restricted, the flow pat- tern in water boxes, leading to excessive inlet erosion of the tubing.

It is important, for good performance of the inlet section tubing, to avoid cast iron or carbon steel surfaces or components, coatings, linings and cathodic protection of internal pipe and water box surfaces in contact with seawater downstream of the screens and settling basin. There is no better way to insure much downtime and frequent tube plugging and replacement than to add rust from steel and cast iron or to add sections of spalled coatings to the fine debris which passes through the screens.

It is quite possible to block the tubes, as shown by Figure 1. Even worse than tube blocking, is the all too common problem of a piece of rust or lining lodging itself in a tube, creating local turbulence and finally causing a tube failure through erosion.

Figure 2 shows a stone partially blocking a tube, and the hole caused by the excessive turbulence downstream. Steel and coatings have their useful and proper place but it is false economy to save money on cheaper pipe and water boxes if their use can lead to tube failure.

When the cost of a good coating or lining, plus the cost of the supplementary cathodic protection system, is added to the first cost of the steel, cast iron or concrete, an all-alloy system is found to be much more competitive in first cost than commonly believed. More important, however, is the major saving that comes with increased on-stream time during the full lifetime of the plant.

The authors believe that when the basic principles of full life cycle economics are properly applied, as they ultimately are applied in every industry, the inlet heat reject section inevitably becomes a nonferrous alloy system.

The water box serves to conduct the water from the piping system to the inlet of the condenser tubes. It must be designed to assure that the dissolved gases, released as the water expands from the pipe into the large water box chamber, are removed by proper venting.

If not properly vented an air lock can develop and block off the upper tubes. This can lead to hotter tubes, hot spots and tube failures. The water box must also be designed to distribute the water from the pipe in a uniform, streamlined fashion, more or less equally to all the tubes, to minimize turbulence at the tube inlets. Quick opening (Tube-Turn type) inspection handholes, epoxy lined or alloy, properly located are a "must" for minimum downtime.

One economical method of lining water boxes with 90-10 copper-nickel sheet is shown in Figure 3. Although 90-10 copper-nickel alloy can be welded directly to steel, the problem in applying thin linings to steel water boxes is to attach them securely to the steel so that they can withstand vacuum conditions should these arise. An economic method of doing this, using M.I.G. spot welding, was developed by International Nickel Limited. This enables spot welds to be made through the sheet, fastening it to the water box without weld preparation other than cleaning the surfaces.

FIGURE 1: TUBES IN DESALINATION PLANT BLOCKED BY DEBRIS

FIGURE 2: 70-30 COPPER-NICKEL TUBE PARTIALLY BLOCKED BY STONE CAUSING TURBULENCE AND PERFORATION DOWNSTREAM

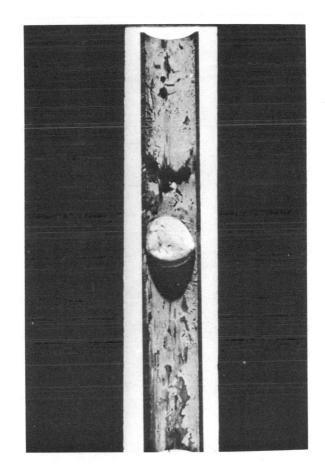

FIGURE 3: SPOT WELDING A THIN LINING OF CUPRO-NICKEL TO A WATER BOX

A 70-30 copper-nickel alloy filter metal containing titanium is used to make the weld, as this serves to reduce weld porosity and gives a spot weld which is galvanically compatible with the lining. Iron pickup in the weld spot can be controlled and kept to a low level by correct welding techniques.

Seawater Pumps

Pumps are an essential component of a plant and, if faulty, can lead to frequent plant shutdowns (see Table 8). Suction strainers are normally provided on vertical deep well-type seawater pumps and should be supplied on all seawater supply pumps to protect the pump itself. To provide further protection of the heat reject condenser, it is advisable to fit duplex strainers on the discharge of the supply pumps.

To provide the same protection against rust flakes becoming lodged in the heat recovery condensers, a strainer should also be provided on the suction side of the brine recycle pumps. These strainers must be arranged so that they can be cleaned under way while operating with the installed standby recycle pump. Type 316 stainless steel wire mesh strainers have served well to date.

Wear rings, seals and packing are the components requiring most frequent attention and renewal. Since increasing wear ring clearance results in decreasing discharge head, it is important to both pump efficiency and plant on-stream time to use durable wear ring materials.

Although wear rings are generally designed to operate without contact, separated by 10 to 20 mil (0.25 to 0.05 mm.) initial clearance, grooving or galling is observed in many pump installations. To reduce galling, one or both of the rings are often made from cast higher alloyed stainless or nickel base alloys such as CD4M-Cu precipitation hardening stainless steel or even Inconel X-750.

Proper heat treatment is essential for dimensional stability and ductility in all precipitation hardening grades. In many designs, operational clearances are such that the cast nonhardening materials such as CF8M (316), Monel Alloy 506 (H Monel) and CN7M (Alloy 20) have given excellent performance.

The turbulence in the area of the wear rings is such that most copper base alloys and Ni-Resist require frequent replacement if used for wear rings. Seals, packing and sleeves are also subject to wear and replacement.

Typical combinations that have given reasonable service are listed in Table 8. Whereas wear rings can be selected to give several years' continuous service, provision must be made for periodic renewal of either mechanical seals or packing. One face of the mechanical seal is normally carbon, of which there are many grades.

The metal filled grades of carbon suffer more corrosion on seawater pumps and generally require earlier renewal than the phenolic impregnated grades of carbon.

TABLE 8: MATERIALS FOR SEAWATER PUMPS

(A) Vertical Wet Pit-Type Seawater and Brine Recycle Pumps

Class of Service	Basic Minimum Cost Construction[1] Frequent, but predictable replacement	Preferred High service factor Low maintenance Infrequent maintenance	Most Durable Minimum out-of-service time Minimum maintenance
Body			
Suction bell	Cast iron 2% Ni Cast iron	Ni-Resist[2] Type I or Type II 4½% Ni-Al bronze	Ni-Resist Type I or Type II
Shroud, liner or case in way of impeller	Cast iron 2% Ni Cast iron Cement or epoxy-lined cast iron G or M bronze	CF3M (316L) 4½% Ni-Al bronze	CF3M (316L)
Case with straightening vanes (diffuser)	Cast iron 2% Ni Cast iron Cement or epoxy-lined cast iron G or M bronze	Ni-Resist Type Ib[3] 4½% Ni-Al bronze CF3M (316L)	CF3M (316L)
Discharge column	Cast iron 2% Ni Cast iron Carbon steel	Cast iron 2% Ni Cast iron Cement or epoxy-lined cast iron or steel CF3M (316L) 90-10 Cu-Ni Ni-Resist Type I or II	CF3M (316L) 90-10 Cu-Ni Ni-Resist Type I or II

(continued)

TABLE 8: (continued)

	Basic Minimum Cost Construction[1]	Preferred	Most Durable
Deepwell can (brine & recycle pumps)	Carbon steel, cement or epoxy-lined	Ni-Resist Type I or II (impregnated for vacuum service) 4½% Ni-Al bronze 90-10 Cu-Ni	90-10 Cu-Ni
Impeller	G or M bronze	4½% Ni-Al bronze[4] CF3M (304L)[5] CF8M (316) 29 Cr-11 Ni	CF8M (316) 29 Cr-11 Ni
Wear Rings	G or M bronze Ni-Resist Type I or II	Type 316 CN7M (Alloy 20) CD4M-Cu Monel* Alloy 506 (H Monel) Inconel* X-750	Monel* Alloy 506 (H Monel) Inconel* X-750
Shaft	Type 410 Precipitation-hardening Stainless steel	Type 316 Mone* Alloy 400 Monel* Alloy 500	Type 316 Monel* Alloy 400 Monel* Alloy K500
Bolting	Silicon bronze	Monel* Alloy 400	Monel* Alloy 400

[1]Common cast iron-bronze trimmed pumps will pump seawater and are widely used for temporary installations and even permanent installations where high maintenance and frequent out-of-service periods are tolerable. Brasses (copper-base alloys containing zinc) are unsatisfactory and corrode sacrificially to other components. They are not listed, even under minimum cost construction.

[2]Ni-Resist refers to a family of austinitic nickel cast irons. See ASTM Spec. A 436.

(continued)

TABLE 8: (continued)

[3]The higher chromium grade of Ni-Resist, Type Ib, is preferred to better resist severe turbulence in the way of straightening vanes if design of this section leads itself to the higher shrinkage rates encountered as chromium content is increased. If not, upgrade to CF3M (316L).

[4]Aluminum bronze dezincifies rapidly, even in quiet seawater, with little visible indication of attack until the dezincified shell collapses. The use of grades with 4½% minimum nickel avoids dezincification. Check with supplier before weld repairs are made, since even with 4½% nickel, dezincification may be encountered in the heat-affected zone of the weld. Nickel aluminum bronze has given excellent service in pump impellers, although its use has been limited. The stainless alloys have an advantage under the most severe velocity conditions, but in many designs nickel aluminum bronze could approach the durability of stainless and it is more resistant to pitting during down periods.

[5]CF3 has given excellent performance as workboat propellers with favorable but limited use in pump impellers. The molybdenum-containing grade has better resistance to pitting in seawater and provides a better margin of safety during down periods.

*Trademark, The International Nickel Company, Inc.

(B) Volute Dry Pit-Type Seawater and Brine Recycle Pumps

Class of Service	Basic Minimum Construction[1]	Preferred For small and medium pumps; shipboard service[2]	Preferred Coastal plants[3]	Most Durable
Body	Cast iron Nickel cast iron	G or M Bronze 4½% Ni-Al bronze 70-30 Cu-Ni (weldable)	Ni-Resist Type I, Ib or II	
Wear Ring	G or M bronze		Type 316 CN7M (Alloy 20) CD4M-Cu	Monel* Alloy 506 (H Monel) Inconel X-750
Shaft	Bronze	Monel* Alloy 400 Monel* Alloy K500	Type 316 Monel*Alloy 400	Monel* Alloy K500

(continued)

TABLE 8: (continued)

	Basic Minimum Construction[1]	Preferred	Preferred	Most Durable
Impeller	G or M Bronze	4½% Ni-Al bronze / Monel* Alloy 400 (A Monel) / Monel* Alloy 411 (E Monel)	4½% Ni-Al bronze / CF8M (316)	Worthite** / 29 Cr-11 Ni
Bolting	Carbon Steel	Monel* Alloy 400	Monel* Alloy 400	

(C) Special Seawater Pump Constructions

Class of Service	Preferred		Most Durable	
	A[4]	B[5]	C[6]	D[6]
Body	CF3M (316L)	4½% Ni-Al bronze	Worthite**	CN7M
Wear Ring	Type 316	Monel* Alloy 506	Worthite** / Inconel* X-750	Alloy 20 / Inconel* X-750
Shaft	Type 316	Monel* Alloy 400	Alloy 20	Alloy 20
Impeller	CF8M (316)	4½% Ni-Al bronze	Worthite**	CN7M
Bolting	Type 316	Monel* Alloy 400	Type 316 / Monel* Alloy 400	Type 316 / Monel* Alloy 400

[1] Lower service factor. Frequent but scheduled replacement.

[2] Bronze cases with Monel Alloy 410 impellers should be considered good standard construction for merchant ships and small mass produced pumps. The U.S. Navy has shifted from G and M bronze to 70-30 Cu-Ni cases in order to facilitate repair by welding.

(continued)

TABLE 8: (continued)

[3]Ni-Resist provides cathodic protection for CF8M impellers and Type 316 shafts that tend to minimize pitting and crevice corrosion during downtime.

[4]All CF3M (316L) seawater pumps are subject to crevice corrosion and pitting unless considerable care is taken to open the pumps and wash down with fresh water whenever they are removed from service.

[5]The bronze body will provide adequate cathodic protection to internal Monel components to essentially eliminate crevice corrosion of these components during down periods. Unfortunately, bronze is not as effective in preventing crevice corrosion of CF8M (Type 316) internal components.

[6]Higher alloyed pumps such as Worthite and Alloy 20 are substantially more resistant to crevice corrosion and pitting. Limited experience indicates it may not be necessary to clean or wash down with fresh water as frequently as with the all CF8M (316) construction.

*Trademark, The International Nickel Company, Inc.

**Trademark, Worthington Corporation

Since metals are anodic to carbon, it is remarkable that any metal can be operated in contact with carbon without suffering aggravated galvanic corrosion. The authors can only conclude that the carbon in some grades of phenolic impregnated carbons is in a form that is less active galvanically than one would anticipate.

Certainly the graphite lubricated packings, inadvertently used in pump and valve stem packings, have caused premature failures of sleeves and shafts and should be avoided unless high sleeve and shaft replacement rates are acceptable. The complexities of the seal, packing and sleeve problems are such that the manufacturer's actual seawater operating experience should be sought and evaluated.

Impellers generally outlast wear rings and seals but may be replaced if they show signs of wear when the rings and seals are renewed. Cast stainless steels CF3 (304L), CF8M (316) and many special compositions such as 29 Cr-11 Ni have given more than 10 years good performance in cast iron, steel or Ni-Resist cases where they receive enough cathodic protection to overcome their tendency toward crevice corrosion, particularly during idle periods.

In bronze-bodied pumps, cast Monel Alloy 506 impellers give good performance, since the case again protects the impeller from crevice corrosion. Impellers and casings of the propeller alloy, 4½% nickel-aluminum bronze, have given excellent service.

The higher alloyed steels such as Worthite and Alloy 20 (CN7M) have sufficient resistance to crevice corrosion to give generally reliable performance as impellers in a seawater pump, regardless of the case material. In the smaller sizes, all Worthite or all Alloy 20 (CN7M) pumps represent an economic choice in terms of minimum-cost, maximum trouble-free performance.

The casing generally requires less replacement than impellers unless low first cost induces the owner to accept carbon steel or cast iron with cement or other linings. Steel or cast iron pump bodies may represent economy in areas with good repair and maintenance facilities but not elsewhere. Tin bronze, aluminum bronze and Ni-Resist cases perform well in pumps where care is taken in the design to protect the case from the worst turbulence with a thick enough cushion of more slowly moving water. In many vertical pump designs the casing in the way of the impeller is sleeved with, or made from, CF8M (316) stainless steel.

The Office of Saline Water is evaluating large, all-stainless Type CF8M (316) seawater pumps, with carbon steel anodes built into specially designed recesses in the case to reduce crevice corrosion. All stainless steel pumps can be used successfully if they are carefully flushed out with fresh water and drained when shut down.

Otherwise they may suffer severe crevice corrosion while on standby, unless galvanically protected as in the case of the specifically designed O.S.W. pumps.

Seawater Valves

The rubber seated butterfly valve would seem to have been designed and developed with desalination plants in mind. Details of materials are shown in Table 9. Those designs where the rubber seat fully covers the body would seem at first to permit the use of lower cost carbon steel or ductile iron bodies.

However, experience has shown that seawater seeps through the rubber seat to valve stem joint. Rusting of steel and iron bodied valves is followed by enough buildup of corrosion products to raise the rubber seat slightly and interfere with proper closing of the valve.

TABLE 9: SEAWATER VALVES

Butterfly Valve

Valve Body
    Cast iron, S.G. iron, carbon steel. Protected by elastomeric lining.[*]
    Ni-Resist irons Types 1 and 2.
    Gunmetal e.g., 88-10-2 copper-tin-zinc.
    4½% nickel-aluminum bronze.
    Type 316 may be subject to pitting and crevice corrosion particularly
       in dead-ended lines and in closed position.
    Cast nickel-copper alloys — may pit in same way as Type 316, but
       pitting tends to be shallow.

Valve Disc
    Ni-Resist irons Types 1 and 2.
    4½% nickel-aluminum bronze.
    Cast nickel-copper alloys.

Valve Spindle
    Monel Alloy K500.
    Monel Alloy 400.
    Stainless steel Type 316 — may suffer crevice corrosion under packing.

Packing
    Graphite lubricated types can cause galvanic corrosion of the spindle.
       Packing Teflon is suitable for butterfly valve.

Gate and Globe Valves

There are such a large number and variety of cast and wrought alloys used by the many manufacturers of gate and globe valves that a separate paper is required to treat the subject properly. Until such a paper is prepared, Table 8 — Materials for Seawater Pumps — can be used to establish general guidelines. Seats and discs are subject to much the same turbulence as wear rings and impellers. Valve stems and pump shafts are comparable. Valve bodies and pump bodies are also roughly comparable. Galvanic compatability and turbulent resistant seats and discs are key considerations in obtaining reasonable service and low replacement rates in gate and globe valves.

[*]Weld deposit of Alloy 625 or baked epoxy under the elastomeric lining is necessary to prevent scale buildup under the lining and interference with closure.

For smaller diameter valves, Ni-Resist or nickel-aluminum bronze (10% Al, 5% Ni, 5% Fe) bodies are suggested. For larger valves, where solid alloy would not be economical, the body can be overlaid with Inconel Alloy 625 or coated with epoxy beneath the rubber lining.

The 4½% nickel-aluminum bronze is widely used and appears to be the preferred material for the disc in valves of moderate size for the disc, particularly in valves larger than 24 inches, though its lower tolerance for velocity effects has led, in some installations, to undesirable roughening of seat surface. This has resulted in some experimentation with stainless, nickel and Monel overlay on the seat face, particularly in butterfly valves used as control valves.

Acid Treatment and Deaeration

Another area of the desalination plant where materials selection is critical to performance and higher on-stream time is the acid treatment and deaeration section, between the outlet of the heat reject section and the point of make up addition in the recirculating brine stream. 90-10 copper-nickel piping serves well up to the point of acid injection

At the point of acid injection and for several diameters downstream, fiberglass reinforced epoxy and fiberglass reinforced PVC lined piping have been found useful when suitably protected from ultraviolet rays of sunlight. The 90-10 copper-nickel piping serves well in this section only if the acid mixing is thorough and the intended pH of 4.0 or 4.5 is reached uniformly within ½ to 1 diameters downstream of the acid injection point. With less complete mixing, slugs of very low pH acid can reach the pipe wall several diameters downstream and give rise to excessive corrosion.

The dissolved gases vented and stripped from the warmed seawater in the deaerator include oxygen from normal seawater and considerable volumes of CO_2 from the reaction of the injected acid with carbonate. These highly corrosive wet gases should be removed in a separate deaerator vessel (not designed as a part of the flash evaporator) and vented to a relatively small vent condenser, constructed of special alloy materials, prior to passing into the ejector system.

In many coastal waters pollution can lead to substantial H_2S in the gas system. The use of 70-30 copper-nickel alloy or even Monel Alloy 400 for the shell of the deaerator and the tubing of the deaerator vent condensor is marginal at best and dependent upon the actual amount of CO_2 and O_2 in the gas stream.

These alloys require frequent replacement if pollutants introduce substantial H_2S in the gas stream being vented and stripped in the deaerator. Type 316 stainless steel, Carpenter 20Cb-3, Incoloy 825 and even titanium may be required for the shell and tubes of the deaeration section if the gas stream is foul enough to corrode 70-30 copper-nickel or Monel Alloy 400.

It is economical of course to use the more highly alloyed materials in the relatively small portion of the plant represented by the deaerator where these corrosive gases are stripped out to reduce the corrosiveness of the seawater to the main section of the plant downstream of the deaerator. One or two plants have tried acid injection without a deaeration section, only to find that carbon steel chamber walls corrode away in a few years.

The degree to which the deaerator strips oxygen, CO_2 and H_2S from the seawater is directly related to the rate of corrosion of the chamber walls and other steel components in the main portion of the plant and is a key factor in determining whether a given plant will survive and produce water economically for 10, 20 or 30 years.

Brine Heater

Steam normally provides the heat input for the whole plant in the relatively small brine heater. The problem here is scaling, a problem that affects the design and daily operation of the whole plant. If the plant is designed and operated so that the scale deposition does not occur or occurs very slowly, corrosion in the brine heater does not appear to be a serious problem.

Uneven scale deposition from operational upsets or nonuniform distribution of the steam supplying the heat has led to hot spot corrosion of the brine heater tubing in a few plants. The conditions that lead to corrosion of brine heater tubing must be eliminated or brought under control if the plant is to work economically. Aluminum brass, 70-30 copper-nickel and 90-10 copper-nickel, especially in the plants using acid treatment and pushing brine heater temperatures as close to the scaling temperature as practical.

Product Water

The product water, which is condensed vapor above the brine in each chamber or stage, is collected on trays below the tubing and drained through piping to the product water pump and product water treatment tank. The product water is corrosive to carbon steel and should be handled in stainless steel (304L) until it reaches the product water treatment tank.

Instrument Cases, Gauges and Actuator Linkages

A walk around any desalination plant will reveal moderate to severe deterioration of painted steel gauges, instrument cases, electrical switch boxes, handrails, ladders and even the linkages of motor or hydraulic operated valves.

Careful attention to original selection of materials for the components can do much to reduce "routine" maintenance, particularly during the later stages of a plant's normal lifetime.

Thin wall stainless steel handrails, ladders and stainless steel gauge and instrument cases can often be obtained for little more than the less durable painted steel items they replace if the owner insists upon and seeks suppliers furnishing these items in stainless steel.

CONCLUSION

From the seawater intake to the potable water delivery pipe, conditions in a distillation plant pose a variety of complex problems in materials selections as well as in plant design and control. A growing body of knowledge relating to laboratory data, inplant tests and operating experience suggests criteria for proper selection.

Apart from estimating initial cost of the mill forms which are available, account has to be taken of fabrication costs and service performance. While ability to carry out repairs and the cost of repairs are important, the loss of product itself and the serious consequences of having an unreliable source of an essential supply are of greater concern. It has been shown that by the selection of materials of proven performance, service between overhauls can be extended and unscheduled stops minimized.

Gate and Globe Valves

There are such a large number and variety of cast and wrought alloys used by the many manufacturers of gate and globe valves that a separate paper is required to treat the subject properly. Until such a paper is prepared, Table 8 — Materials for Seawater Pumps — can be used to establish general guidelines.

Seats and discs are subject to much the same turbulence as wear rings and impellers. Valve stems and pump shafts are comparable. Valve bodies and pump bodies are also roughly comparable. Galvanic compatability and turbulent resistant seats and discs are key considerations in obtaining reasonable service and low replacement rates in gate and globe valves.

REFERENCES

General

U.S. Dept. of the Interior, O.S.W., Washington, D.C., "Annual Saline Water Conversion Reports".

Nace, R.L., "Water of the World", Natural History, 73, #1, (January 1964), pp. 10-19.

Improved Heat Transfer

Alexander, L.G., Joyne, J.D. and Hoffman, H.W., Advanced VTE Heat Transfer Surfaces.

Withers, J.G. and Young, E.H., Investigation of Steam Condensing on Vertical Rows of Horizontal Corrugated and Plain Tubes.

Erb, R.A., Haigh, T.I., and Downing, T.M., Permanent Systems for Dropwise Condensation for Distillation Plants.

Ford, J.A. and Butt, S.H., Helically Formed Enhanced Exchanger Tube.

Campbell, K.S. and Lennox, J.P., Large Plant Performance Testing of a Commercial Size Double Fluted Tube Bundle.

Kays, D.D., Application on Enhanced Heat Transfer Surfaces to Vertical Tube Evaporators.

Performance of Materials

Bom, P.R., Consultant, Werkspoor-Amsterdam, NV, Selection of Alloys for Multi-Stage Flash-Distillation Plant.

Hunt, J.R. and Bellware, M.D., "Ocean Engineering Hardware Requires Copper-Nickel Alloys." Trans. of Third Annual MTS Conference & Exhibit, June 5-7, 1967, pp. 243-275.

Tuthill, A.H. and Sudrabin, D.A., "Why Copper-Nickel Alloys for Desalination." Paper before ASM, Natl. Exposition Congress Chicago, October 31 to November 3, 1966, p. 36; Metals Eng. Quart, Vol. 7, #3, (August 1967), pp. 10-26.

Brown, B.J., "Metals and Corrosion." Machine Design, Vol. 40, #2 (Jan. 18, 1969), pp. 165-173.

U.S. Dept. of the Interior, O.S.W., Washington, D.C., Annual Reports, Seawater Desalting Demonstration Plant #1, Freeport, Texas.

"Galvanic Corrosion Shuts Down Freeport's Salt Water Plant" Corrosion, Vol. 17, (September 1969), p. 37.

Tuthill, A.H. and Weldon, B., "The Copper-Nickels in Desalination Plants", U.K. Copper Development Association, Desalination Conference, Dec. 8, 1966.

Hunter, J.A., Director, "Indexed Bibliography on Corrosion and Performance of Materials in Saline Water Conversion Process — 1, 2, 3, 4", U.S. Dept. of the Interior O.S.W., Oak Ridge Nat'l. Lab., December 1968.

Moore, Robert E., "Materials for Water Desalting Plants." Chemical Engineering, Vol. 70, #20, (September 30, 1963), pp. 124-128. Part II, Vol. 70, #21, (October 14, 1963), pp. 224-230.

Todd, B., "The Corrosion of Materials in Desalination Plants." Second European Symposium on Fresh Water from the Sea, Athens, May 9-12, 1967; Desalination, Vol. 3, #1, (1967), pp. 106-117.

Bell, W.S. and Bramer, H.C., Cyrus William Rice & Co. for O.S.W., U.S. "Materials Evaluation Program." Research & Dev. Progress Report No. 308.

Carbon Steel

Shea, E.P., Hollingshad, H.C., Cyrus William Rice & Co. for O.S.W., "Corrosion in Carbon Steel Heat Transfer Surfaces," Report No. 166, December 1965.

Konstantinova, E.V. and Semenova, L.S., "Corrosion of Mild Steel Under the Operating Conditions of the Distillation Desalting Plant," Second European Symposium on Fresh Water from the Sea, Athens, May 9-12, 1967 Desalination, Vol. 5, #1, (1968), pp. 90-94.

Concrete

Higginson, E.C. and Backstrom, J.E., Bureau of Reclamation, U.S. Dept. of Interior, "Evaluation of Concrete for Desalination Plants." Journal of Structural Div., ASCE Water Resources Engineering Conference, Denver, Colorado, May 16-20, 1966, Feb. 1968.

Pumps

U.S. Dept. of the Interior, O.S.W., "Research on Pumping Unit Studies Final Evaluation Report." Research & Development Report No. 205, Sept. 1966.

Chlorination

Anderson, D.B. and Richards, B.R., "Chlorination of Sea Water — Effects on Fouling and Corrosion," Trans. ASME, Ser. A., J. Eng. for Power, Vol. 88, #3, (July 1966), pp. 203-208.

Linings

Todhunter, Harold A., "Special Linings in Concrete Intakes Prevent Fouling from Marine Growth." Power Engineering Magazine, Vol. 70, #11, (November 1966), p. 67.

Todhunter, Harold A., "Condenser Waterbox Service." Materials Protection, Vol. 6, #7, (July 1967), pp. 45-46.

Ridgway, W.F. and Heath, D.J., "Lining Mild Steel Components with 90-10 Copper-Nickel Alloy Sheet." Welding & Metal Fabrication, October 1969.

Orde, P.K., "Desalination — A Critical Review of its Application." Journal of the Royal Society of Arts, August 1969, pp. 672-673.

Fasteners

"Fastener Guide for the Marine Industry." Technical Bulletin #109, H.M., Harper Co., 1967.

Bladholm, E.F., "Stainless Steel in Coastal Power Plants." Materials Protection, August 1962, p. 32.

Tubing

Newton, Emerson H., "Survey of Condenser Tube Life in Salt Water Service." Arthur D. Little Inc., International Nickel Power Conference, May 21 and 24, 1968.

Newton, E.H. and Birkett, J.D. Summary Report on "Survey of Materials Behaviour in Multi-Stage Flash Distillation Plants." Arthur D. Little, Inc., O.S.W. Symposium, Washington, D.C., September 25, 1968.

Al Bronze

Niederberger, R.B., "Composition and Heat Treatment Effect on Dealuminization of Aluminum Bronzes." Trans. AFS 72, 1964, pp. 115-128. Modern Castings, Vol. 45, #3, (March 1964), pp. 115-128, (New Technol. Sheet).

Operating Experience

Tidball, R.A., Gaydos, J.G., King, W.M., "Operating Experiences of One MDG Desalination Plant on the Red Sea." Presented at Western Water and Power Symposium, Los Angeles, Calif., April 8 and 9, 1968, Proceedings, pp. C-43-49.

Mulford, Stewart F., "The San Diego Test Facility." Western Water & Power Symposium, Los Angeles, April 8, 1968.

Arad, N., Mulford, S.F., Wilson, J.R., "Desalting Plant Down Time Predicted by Formula." Environment Science and Technology, Vol. 2, #6, (June 1968), pp. 420-427.

Hammond, R.P., Oak Ridge Natl. Lab. "Deaerators for Desalination Plants." R & D Progress Report No. 314.

Arad N., Mulford, S.F. & Wilson, J.R., "Prediction of Large Desalting Plant Availability Factor." Second European Symposium on Fresh Water from the Sea, Athens, May 9-12, 1967; Desalination, Vol. 3, #1, (1967), pp. 378-383.

El-Saie, M.H.A., "The Latest Five Million Imperial Gallons Per Day Flash Distillation Plant in Kuwait." Second European Symposium on Fresh Water from the Sea, Athens, May 9-12, 1967.

THE USE OF NONMETALLICS IN
DISTILLATION TYPE DESALINATION PLANTS

Oliver Osborn
The Dow Chemical Company
Freeport, Texas

ABSTRACT

The next generation of desalination plants will contain a high percentage of
nonmetallic materials. The properties, advantages, and disadvantages of the
various classes of plastics, as well as physical and economic comparisons with
metallic materials, have been listed.

For several years the Office of Saline Water, U.S. Department of the Interior,
has pursued a materials development program for applications of nonmetallics
in desalination plants. A summary of their reported work on these applications
is included in this paper. Current applications in desalination plants today are
covered, with illustrations from the OSW Freeport Test Facility in Freeport,
Texas.

INTRODUCTION

A good deal of visionary projection can be employed in a discussion of the use
of nonmetallics in desalination plants, since nonmetallics are in a much less advanced stage of development than are metals. However, many concrete facts
on this subject are also available.

The basic question is: "Why consider nonmetallic materials at all in the construction of desalination plants?" In many cases in which the plant builder
must pay a premium for metallics with added chemical resistance, it appears
that he will get a better bargain with nonmetallic materials. "If this statement
is true," you may ask, "why aren't we building all nonmetallic desalination
plants?" Here again, the simple answer is that no one knows how long they
are going to last.

It is my prediction that the next generation of desalination plants will contain a high percentage of nonmetallics; in fact, this generation will represent a wedding between the metals and the nonmetals.

Unlike the metals industry, the plastics industry has not yet united to exploit the use of plastics in desalination. Therefore, the Office of Saline Water (OSW) has initiated an extensive nonmetallics development and evaluation program. A discussion of the basic properties and economics of the various families of plastics which are under development by industry will be presented in this paper. This will include both the advantages and disadvantages of these materials as well as a description of the applications of nonmetallics currently employed in Office of Saline Water desalination plants.

Following are some of the specific places in which nonmetallics already fit, or may ultimately fit, in the complex layout of the typical desalination plant (Figure 1).

FIGURE 1: ARTIST'S CONCEPTION OF TYPICAL DESALINATION PLANT

The intake system employs a "grizzly" that keeps large trash from entering the plant. This grizzly may be fabricated of a fiber glass reinforced plastic (FRP) grillwork. The intake basin may be constructed of concrete which is, of course, a nonmetallic. The structural members of the intake screens may be plastic coated steel, and ultimately the screens may be made entirely of plastic. The intake pumps may very well have epoxy coated linings.

In the water pretreatment system, the flocculation tank, or settling tank, will probably have a painted coating on it to protect it from corrosion. The lines that carry the water from the intake to the plant may etiher be fiber glass reinforced plastic, concrete or painted steel.

In the softening plant, all the vessels and the lines may be made from fiber glass reinforced plastic. The structure in which these vessels are contained may also be made of this material. The grillwork around the plant may be FRP grating just as the intake grizzlies were. The softened water storage tank may be a basin fashioned in the soil which has a plastic film liner.

The deaerator system, like the softening plant, may include vessels and lines constructed of FRP material. The packing in the deaerator tower may be polyethylene or polypropylene. The caustic lines and perhaps the acid lines (if concentrated sulfuric acid is not being used) may also be of plastic lined steel.

In the Vertical Tube Evaporator (VTE) portion of this plant, the areas in which plastics or nonmetallics will be used may include the transfer lines (either FRP or concrete coated), and the bodies of the evaporators (concrete, concrete lined or possibly plastic lined steel). The vertical tubes from which the evaporation is taking place may have a thin plastic film on the outside to promote dropwise condensation.

In the multistage flash distillation section of this hypothetical plant, the evaporation boxes may be fashioned from concrete lined steel or perhaps FRP lined steel, and, of course, all the transfer lines and the gratings may be made from the same material. After the water is produced, it must be stored. Since it is definitely corrosive, FRP tanks may be employed. Also, the effluent from the plant, which is a highly saline brine, may be held in ponds lined with plastic films while being evaporated before disposal. Finally, the return of the effluent to the sea may be accomplished with the use of concrete lined flumes.

STRUCTURAL PLASTICS

It might be beneficial to look at some of the basic properties of various plastic materials. Structural plastics are divided into two classes — the thermoplastic materials (see Table 1) and the thermoset materials. The thermoplastic materials are those which will soften under heat and can be molded into desired shapes. The thermoset materials originate as viscous liquids which turn into a permanent solid when catalyzed or brought to a particular temperature. These thermoset materials have an advantage over the thermoplastic materials in that various reinforcing binders, such as fiber glass, can be incorporated into them before they are set. Thermoplastic materials have less strength than and cannot be used in as many applications as the thermoset materials. Because of this lack of strength, the thermoplastics are not used a great deal industrially unless they are encased in a steel liner.

Table 1 gives a list of the properties of thermoplastics, primarily in an increasing cost sequence. However, both temperature stability and ultimate strength approximately follow this same sequence. Polyethylene is the least expensive material, and polyvinyl dichloride is the most expensive, but also has the best properties. In between, we have acrylonitrile butadiene styrene (ABS) copolymers, polypropylene, cross-linked polyethylene and polyvinyl chloride. These materials are used mainly in the home for hoses, toys, various plastic kitchen articles, and some home and industrial piping.

The thermoset materials are those that would primarily be used in the desalination environments. Their properties are shown in Table 2. The tensile strengths of these materials are much higher than the thermoplastics. The wide spread in tensile strengh is due to the manner in which the fiber glass has been incorporated into the vehicle — whether it was machine wound or hand laid. Note that the thermal stability of the FRP's is much higher than that of the thermoplastics which are in the same price range.

Let's look at the potential advantages of these fiber glass reinforced thermoset resins. Unlike most metals, they are highly resistant to chemicals and certainly resistant to any environment likely to be encountered in desalination plants. They minimize or eliminate any contamination, such as a heavy metal ion pickup. They are lightweight, can be rapidly installed, and in many cases are competitive costwise to install when compared to metal alloy systems or other methods of piping.

They have low thermal expansion, and some of them are thermally resistant at very high temperatures. They have a friction factor that is very favorable, a low gas permeability, and act as an electrical nonconductor or as an insulator against galvanic cells when metals are present. They are good thermal insulators; i.e., they have low thermal conductivity. They have high strength-to-weight ratios and are easily fabricated in large and complex shapes.

However, all the problems are not solved. Because of their lower strength at high temperatures, these thermoset resins need more support spacing than steel does. They char in a fire; they are difficult to join at low temperatures. They have some operating limitations in continuous service at high pressure and temperature; they have less tolerance for error in design, fabrication and field erection. There is a tendency toward stress fatigue and notch sensitivity. They are subject to edge effect corrosion if not properly coated with a resin-rich layer. Where the plastic is cut, moisture can penetrate into the layer and destroy it. Also, plastics do not have marine antifouling characteristics.

TABLE 1: PHYSICAL PROPERTIES OF COMMON THERMOPLASTIC MATERIALS

Material	Tensile Strength psi	Practical Maximum Temperature °F.	Price Range $/lb. (Polymer)
Polyethylene, high density	3,100 - 5,500	120	0.14 - 0.40
ABS (acrylonitrile-butadiene-styrene extrusion)	4,000 - 7,000	180	0.27 - 0.38
Polypropylene	4,300 - 5,500	210	0.18 - 0.37
Polyethylene, cross-linked	1,600 - 4,600	120	-
Polyvinyl chloride (PVC)	5,000 - 9,000	150	0.19 - 0.42
Polyvinyl dichloride (PVDS or CPVC)	7,500 - 9,000	210	0.40
Vinylidene chloride (Saran)	3,000 - 5,000	140	0.30 - 0.40
Chlorinated polyether (Penton)	6,000	250	3.60 - 4.75
Polyvinyl fluoride (VF$_2$)	5,500 - 7,400	230	4.90
Fluorinated ethylene propylene (FEP)	2,700 - 3,700	325	5.60 - 9.60
Polychlorotrifluoroethylene (KEL-F)	4,500 - 6,000	330	4.70 - 9.00
Polytetrafluoroethylene (PTFE)	2,000 - 5,000	400	3.25

TABLE 2: PHYSICAL PROPERTIES OF COMMON THERMOSET RESIN LAMINATES

Laminates	Tensile Strength psi*	Practical Maximum Temperature °F.	Price Range $/lb. (Polymer)
General purpose polyester	10 - 25,000	125	$0.28
Isophthalic polyester	10 - 25,000	150	$0.36
Bisphenol polyester	10-- 25,000	250	$0.42
Hydrogenated bisphenol-A polyester	10 - 25,000	250	$0.43
Chlorinated polyester	10 - 25,000	250	$0.50
Vinylester (Derakane® Dow)	10 - 25,000	250	$0.42
Epoxy	12 - 30,000	300	$0.60

*Tensile strength is largely dependent on reinforcement used: type, amount, construction, etc. Polyesters may range from:

Hand lay up polyesters	-	9,000 to 25,000 psi
Asbestos reinforced pipe	-	14,000 psi
Centrifugal cast	-	23,000 to 33,000 psi
Filament wound	-	80,000 to 120,000 psi

Contractors will invariably paint plastic lines, and once this is done, they are difficult to recognize as plastics. Because of this, maintenance workers sometimes rupture these lines by trying to walk on them, and welders inadvertently puncture them with a torch.

Another probable disadvantage is that these materials will probably age faster then metals, although this is yet to be proven.

COMPARISON OF METALS AND PLASTICS

Table 3 compares the properties of various metals with these reinforced plastics. Carbon steel, for instance, has a density of 2.83, whereas a glass reinforced epoxy has a density of 0.065 or about one-fiftieth as great, and the tensile strength of this material is almost double that of carbon steel. The thermal conductivity (which relates inversely to insulating value) is one-fifteenth as great. The strength-to-weight ratio of fiber glass reinforced epoxy is 1,500 — quite a difference.

One of the ways to get the necessary strength characteristics from thermoplastics is to use them as liner materials for steel pipe. These systems are used quite widely in industry.

Table 4 shows an economic comparison between various plastic lined and metal pipe. Beginning with carbon steel at the ratio of unity, we are looking at three different line sizes: 2, 4 and 6 inch lines. These costs are based on an installation with the normal amount of fittings. Carbon steel is the cheapest material available, followed by, in ascending order, aluminum, 304 stainless, steel lined with Saran, polypropylene lined steel, stainless steel, nickel, glass, steel lined with Kynar and Penton (higher temperature plastics), alloy 20, and, finally, 316 stainless.

The size of the line has a direct bearing on the cost comparisons. In most cases, the larger the line, the greater will be the cost in comparison to mild steel; however, the plastic piping cost does not increase at as rapid a rate as does metal piping. The cost of the fiber glass reinforced piping will lie somewhere between carbon steel and aluminum — perhaps a bit less than carbon steel at the larger line sizes.

Therefore, it would appear that the primary justification for going to lined steel pipes rather than to the FRP materials is for added safety factor in case of fatigue or some fabrication imperfection. These lined materials have double protection as opposed to the FRP. However, one must pay for this double protection.

At the present time, there seems to be no particular justification for using these lined pipes in desalination applications, other than in critical reagent lines such as those used for caustic, acid, or chlorine.

TABLE 3: COMPARATIVE PHYSICAL PROPERTIES OF METALS AND REINFORCED PLASTICS*

	Density lb./in.^{3}	Tensile Strength psi x 10^3	Thermal Conductivity (°F./ft.)	Strength / Weight Ratio, 10^3
Carbon Steel 1020	0.283	66	28.0	230
Stainless Steel 316	0.286	85	9.4	300
Hastelloy C	0.324	80	6.5	250
Aluminum	0.098	12	135	122
Glass mat laminate	0.050	9 - 15	1 - 5	300
Composite structure glass mat woven roving	0.065	12 - 20	1.5	308
Glass reinforced epoxy, filament wound	0.065	100	1.5 - 2.0	1,500

*Room temperature

TABLE 4: RELATIVE INSTALLED COST FACTORS OF LINED PIPE — RATIOS REFERENCED TO FIELD-FABRICATED CARBON STEEL

Piping Material	Pipe Size		
	2 inch	4 inch	6 inch
Carbon steel - SCH 40	1.00	1.00	1.00
Aluminum - SCH 40	1.39	1.52	1.65
304 SS - SCH 5	1.48	1.65	1.74
Saran lined	1.52	1.54	1.61
Polypropylene lined	1.55	1.57	1.65
316 SS - SCH 5	1.66	1.88	2.16
304 SS - SCH 40	1.74	2.28	2.34
Nickel - SCH 5	1.93	2.22	2.39
Glass	1.94	2.06	1.99
Kynar lined	1.99	2.30	2.50
Penton lined	1.99	2.23	2.41
Alloy 20 - SCH 5	2.00	2.33	2.76
316 SS - SCH 40	2.07	2.75	3.20
Monel - SCH 5	2.11	2.42	2.75
TFE-FEP lined	2.38	2.88	3.30
Alloy 20 - SCH 40	2.56	3.62	4.30
Glass lined	2.62	2.47	2.37
Nickel - SCH 40	2.71	3.42	3.75
Inconel 600 - SCH 5	2.82	3.16	3.37
Monel - SCH 40	3.04	3.75	4.21
Inconel 600 - SCH 40	3.46	4.47	4.06
Hastelloy B - SCH 5	6.18	5.76	4.90
Titanium lined	7.25	5.26	4.50
Zirconium lined	8.59	6.51	5.13
Hastelloy B - SCH 40	10.00	10.21	9.18
Tantalum lined	15.62	14.32	12.52

OSW DEVELOPMENT PROGRAM FOR NONMETALLICS

The OSW has had a materials development program for application in desalination plants for several years. Much of the information reported in this paper in regard to the application development in the field of nonmetallics can be found in the Office of Saline Water 1970-1971 Annual Saline Water Conversion Report. Development work has been in progress on applications of nonmetallic materials in the following areas.

Pond Liners

Battelle made a survey of some 150 solar evaporation plants. They found every conceivable material being used as pond liners. Polyethylene is the most economical but has shortcomings in strength and life. Polyvinyl chlorides are intermediate, and butyl rubber is the most expensive. Butyl rubber is very tough and can be adhesively joined.

Coatings for Heat Transfer Tubes

It is a well-known fact that if condensation can be produced in dropwise fashion on the vapor side of heat transfer tubes, the overall heat transfer coefficient will be considerably enhanced. The Franklin Institute Research Laboratories have been conducting a program aimed at developing coatings which will promote such condensation. One class of nonmetallic coatings known as parylenes, which are paraffin-type hydrocarbons, have shown promise for this application.

Other Coatings

In the use of coatings for corrosion protection, Lehigh University finds that a copolymer of acrilonitrile and vinyl pyridine is very good on steel, and the following materials are satisfactory on concrete:

> coal tar modified epoxy
> nitrile phenolic
> neoprene
> fused polyvinyl fluoride

The Franklin Institute has investigated coatings for concrete to prevent vacuum inleakage which would permit the use of concrete as a material of construction for vacuum vessels. Elastomeric materials, such as Hypalon and neoprene, were the best materials; however, they had to be embedded into the concrete. Surface applied materials were not satisfactory.

Concrete-Strengthening Polymers

Cooperation between the OSW, the Bureau of Reclamation, and the Atomic Energy Commission has resulted in a highly significant development in the concrete field. By impregnating concrete with certain nonviscous monomers such

as methyl methacrylate and then irradiating with gamma rays to produce a
polymer structure throughout the pores of the concrete, the result has been an
unexpectedly high increase in strength. As indicated in Table 5, compressive
strength increases in some cases have been as great as 160%. The ratios of
polymers shown here have been optimized. An explanation of the coding is
as follows:

TMPTMA   -   trimethylolpropane trimethacrylate
DAP   -   diallylphthalate
MMA   -   methyl methacrylate

In addition to the improved compressive strength, greatly improved durability
in sulfates, salts and acids has been observed. This appears to be a very startling
discovery and will probably result in a new material of construction for both
saline water plants and the process industries.

TABLE 5: COMPRESSIVE STRENGTH OF POLYMER-IMPREGNATED OSW-TYPE CONCRETE

Monomer	Test Temperature °F.	Mean Polymer Loading %	Mean Compress Strength psi	Strength Increase %
None	76	0	8,846	0
60% S-40% TMPTMA	76	3.99	19,213	117
90% DAP-10% MMA	76	4.82	18,428	108
None	250	0	6,304	0
60% S-40% TMPTMA	250	3.88	16,129	156
90% DAP-10% MMA	250	4.91	16,420	161
None	290	0	6,674	0
60% S-40% TMPTMA	290	3.88	15,844	138
90% DAP-10% MMA	290	4.91	14,898	123

Plastic Heat Exchangers

In 1971, Atomics International reported the development of a plastic heat exchanger, an example of which is shown in Figure 2. These tubes are made of
polyvinyl fluoride, commercially produced by Du Pont as Tedlar, and are fashioned from film by a process which is illustrated in Figure 3. This unit has
successfully completed a 6 month endurance test in 200°F. seawater in the
tank and hot condensate inside the tubes.

It has been projected by Atomics that the capital cost will be considerably
lower when this type of exchanger is employed. The ultimate life of the film
will, of course, determine its overall economics. For a 2 mil thick film, heat

transfer coefficients are in the range of 150 Btu's per hour per square foot per degree Fahrenheit, whereas the metals will generally be in the 500 to 1,000 Btu range.

FIGURE 2: POLYVINYL FLUORIDE TUBES USED IN HEAT EXCHANGER

FIGURE 3: PROCESS EQUIPMENT FOR PRODUCING PLASTIC HEAT EXCHANGER TUBES

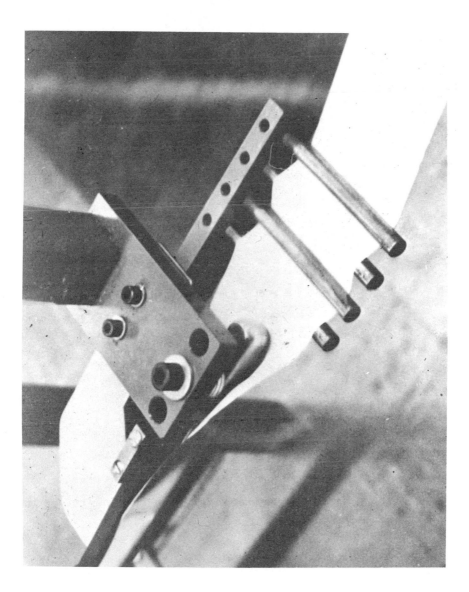

NEW NONMETALLICS MATERIALS EVALUATION FACILITY

For several years, the Dow Chemical Company, under the sponsorship of the
Office of Saline Water, has been testing metallic materials of construction for
use in desalination plants at the OSW Materials Test Center in Freeport, Texas.

Currently, a new nonmetallics materials evaluation facility is being constructed,
the purpose of which is to evaluate changes in physical properties of nonmetallic materials of construction, with time, in flowing seawater at 110°, 160°, 210°
and 250°F., plus steam and concentrated brine, at intervals over a 24 month
period, and to present engineering and physical property changes as a basis for
design data.

Materials which are to be tested are those which are commercially available and
are economically practical, such as fiber glass reinforced plastic systems, thermoplastic systems, paints and coatings, and corrosion-resistant cements.

Properties which are to be measured are summarized into three categories —
plastics, paints and cements — as shown in Table 6.

TABLE 6: PROPERTIES TO BE MEASURED, REPORTED, AND CHANGES
PLOTTED

Plastics	Paints	Cements
Dimensional changes	Blistering	Tensile strength
Weight changes	Cracking	Compressive strength
Tensile strength	Chalking	Flexural strength
Tensile modulus	Cracking	Modulus of elasticity
Flexural strength	Peeling	Abrasion resistance
Flexural modulus	Wrinkling	Absorption
Hardness	Chipping	Porosity/density
Visual-microstructural	Interfilm separation	Quench cycle
Infrared absorption	Film thickness	Maximum temperature
Burst pressure		

Figure 4 is an artist's conception of the test loop. The seawater enters the
unit, is heated to a given temperature, and then is routed through a test cell
which will contain coupons of various shapes and sizes. From the test cell,
it will go into a "trombone" system which will allow the evaluation of various
kinds of piping, both lined and unlined. There will be an individual unit for
each different temperature range to be studied. Such a loop will permit ready
access, removal and replacement of both specimens and piping. The unit is
expected to be in operation within the next few months.

FIGURE 4: ARTIST'S CONCEPTION OF NONMETALLICS TEST LOOP

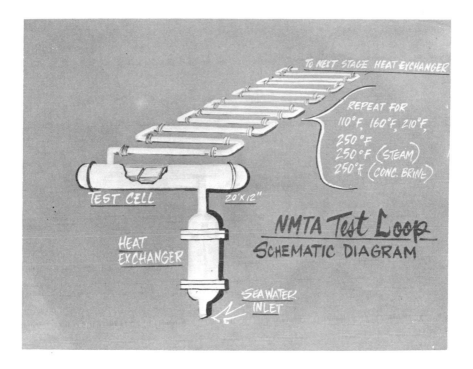

CURRENT APPLICATIONS

There are many applications of nonmetallics in desalination plants today. The OSW Materials Test Center in Freeport, Texas, is the site of the materials evaluation programs already described to you. Figure 5 shows the water treating plant that supplies the feed water for these systems.

A great many nonmetallics were used in this plant for two reasons: (1) to find out how the nonmetallics would perform, and (2) to assure that the water that reached the various metal test loops was not contaminated with any heavy metal ions as it passed through the treating system. Therefore, the acidification and deaeration are carried out almost entirely in nonmetallic vessels or lines. At the top of the photograph (Figure 5) is the deaerator tower. It is a steel tower, lined with a ployamide epoxy coating and loaded with Maspac polypropylene packing. These plastics appear to be holding up very well.

Prior to deaeration, the water is acidified. The acidification is carried out in a fiber glass reinforced vinyl ester tank. Following acidification, there must be

FIGURE 5: WATER TREATING PLANT AT OSW MATERIALS TEST
CENTER, FREEPORT, TEXAS

FIGURE 6: SEAWATER SOFTENING PLANT AT OSW MATERIALS TEST
CENTER, FREEPORT, TEXAS UTILIZING ION EXCHANGE PROCESS

neutralization. This is done by adding caustic to the system, and the caustic storage tank is fabricated of a fiber glass reinforced vinyl ester resin known as Derakane. The steel piping which connects these vessels is lined with either Saran or polypropylene. After the water has been treated, it goes into another fiber glass reinforced vinyl ester storage tank for holding until use.

When this system was designed, it was feared that there would be air inleakage through the plastic wall of this tank and this would prevent the maintaining of the treated water at an oxygen content of essentially zero. However, there was not enough migration of oxygen through the walls to change the oxygen content of the water, even in the 5 to 10 part per billion range.

Another unit of the Materials Test Center in which plastics are used is a seawater softener utilizing an ion exchange process (see Figure 6). This is an invention of Dr. Irwin Higgins, Vice-President of the Chemical Separations Corporation. This unit is in the shape of a huge U-tube through which ion exchange resin is passed. This entire system, excluding the deaerator tower, is fabricated of the fiber glass reinforced vinyl ester to prevent any metal ion contamination. Some of the connecting piping (at lower temperature) is polyvinyl chloride. This is an unreinforced thermoplastic material which has, so far, held up satisfactorily.

Another Materials Test Center structure which has a particularly interesting non-metallic feature is the Stainless Steel Test Plant, shown in Figure 7. The coating on the structural material holding the towers in place is an example of the union of nonmetallics and metallics. This coating is a paint in which stainless steel dust has been intermixed with an epoxy. After two years of service, this paint is essentially in perfect condition.

Figure 8 shows a titanium-clad brine heater operating at 250°F. in one of the experimental materials test units. This heater failed in the head because of a pinhole in a weld. As a quick and economical remedial action, a replacement steel head was fabricated and coated with a forced-cured epoxy system, Plasite 7132/7155. Over the past 90 days, the lining has remained in perfect shape.

Figure 9 presents an interesting plastics application. As you know, most desalination plants are located in extremely corrosive atmospheres because of salt laden atmosphere. This FRP dome has been used quite widely throughout all of the units to protect the various sensitive elements of instrumentation. The pH-measuring equipment, for example, is under this dome.

Let us now move on to the Freeport OSW Test Bed. The current design for this test bed is a union between the multistage flash and the vertical tube-evaporation systems to demonstrate that this combination holds advantages over either system individually. The various places in which nonmetallics are being employed are as follows.

FIGURE 7: STAINLESS STEEL TEST PLANT, MATERIALS TEST CENTER,
FREEPORT, TEXAS

Intake Screens — These traveling screens are constructed of steel coated with
coal tar epoxy, and the structure is encased in a concrete sump (Figure 10).

Seawater Intake Pumps — Figure 11 illustrates the field application of an epoxy
to this type of pump. This is a Dow pump, rather than OSW, but represents a
very successful application of plastics.

Deaerator — The deaerator of the Freeport Plant is fabricated of steel but is
internally coated with an epoxy. The tower is filled with a polypropylene
packing called Maspac (Figure 12).

Stair Treads — These treads are made of a very interesting material called FRP
grating. It is very durable and very corrosion resistant. However, it is about
twice as expensive as galvanized steel.

FIGURE 8: TITANIUM-CLAD BRINE HEATER

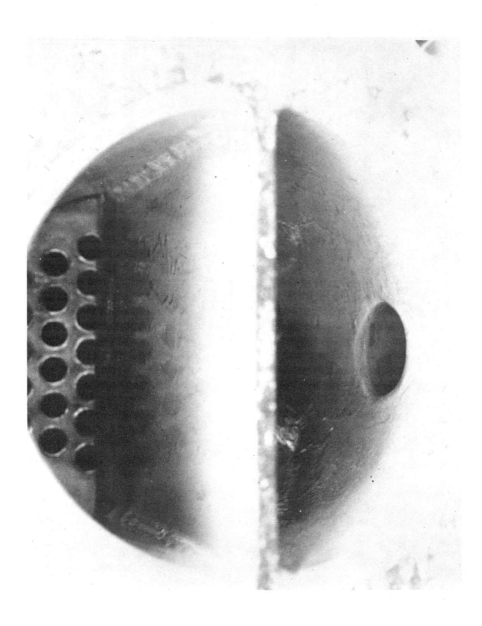

FIGURE 9:   FIBERGLASS REINFORCED PLASTIC DOME TO PROTECT
pH-MEASURING EQUIPMENT

FIGURE 10:   STEEL INTAKE SCREENS COATED WITH COAL TAR-EPOXY
AND ENCASED IN CONCRETE SUMP

FIGURE 11:   FIELD APPLICATION OF EPOXY TO SEAWATER INTAKE
PUMP

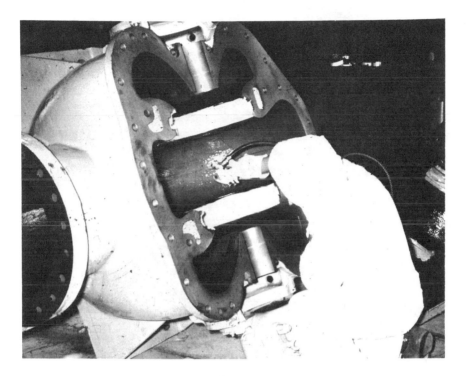

Brine-Carrying Lines — Figure 13 is a view of an 8 inch pipe which is a filament wound, fiber glass reinforced epoxy system that is carrying the brine from the bottom of the MSF system to the brine heater. This line is failing in service as evidenced by the salt crystals at the joint between the pipe and the flange. This failure may have been due to incorrect makeup of the flange. Excessive vibration and lack of support may also be factors in these failures. Flanges of improved design have been installed.

Another one of the transfer lines in the LTV system is shown in Figure 14. This is an experiment to test the durability of concrete-lined pipe which meets the standard AWWS specifications when used at a temperature in the range of 235°F. It is beginning to fail after a year's service.

Steam Chest Coating — Figure 15 is an example of a misapplication of a non-metallic. This is an epoxy coal tar coating that has peeled off in the steam chest of one of the VTE evaporators where the temperature was about 200°F. However, this particular coating was only designed for abut 140°F. This illustrates the ease with which nonmetallics can be misapplied.

FIGURE 12: DEAERATOR TOWER AT FREEPORT OSW TEST BED

Coated internally with epoxy and
filled with polypropylene packing.

FIGURE 13: EIGHT INCH, FILAMENT WOUND, FIBER GLASS REINFORCED EPOXY PIPE

This pipe carries brine to brine heater
at Freeport OSW Test Bed.

FIGURE 14:   CONCRETE-LINED TRANSFER LINE IN LTV SYSTEM OF
FREEPORT OSW TEST BED

FIGURE 15: EPOXY-COAL TAR COATING PEELED FROM STEAM CHEST
IN VTE EVAPORATOR SHOWING MISAPPLICATION
OF A NONMETALLIC

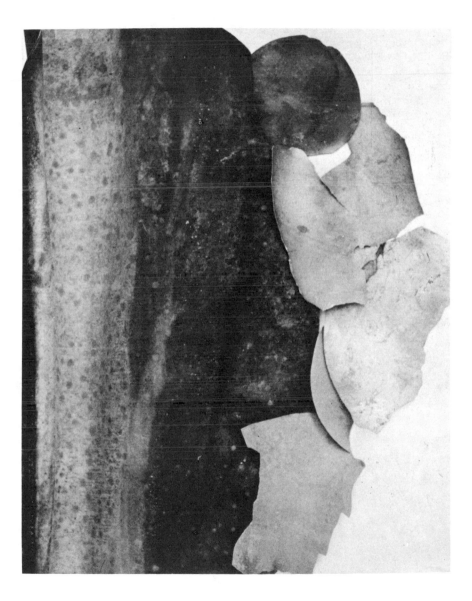

VTE Domes — Figure 16 illustrates a very interesting and very good application for FRP's. These are the VTE domes fabricated from hand-laid, fiber glass reinforced epoxies which are holding up very well. Figure 17 shows a polypropylene distributor weir that fits in the top of one of the VTE tubes to give the water a swirling effect as it enters the tube.

Evaporator Lining — The VTE evaporator is a large steel-shelled vessel which is subject to both liquid and vapor zone corrosion and, if some remedial action were not taken, would fail in two to three years. One of the actions which has been taken at the Freeport Plant is to line these vessels with a steel reinforced special concrete called Prekrete (Figure 18). Over a several year period, this material has been shown to be very durable.

FIGURE 16: VERTICAL TUBE EVAPORATOR DOMES FABRICATED FROM HAND-LAID, FIBER GLASS REINFORCED EPOXIES AT FREEPORT OSW TEST BED

FIGURE 17: POLYPROPYLENE DISTRIBUTOR WEIR

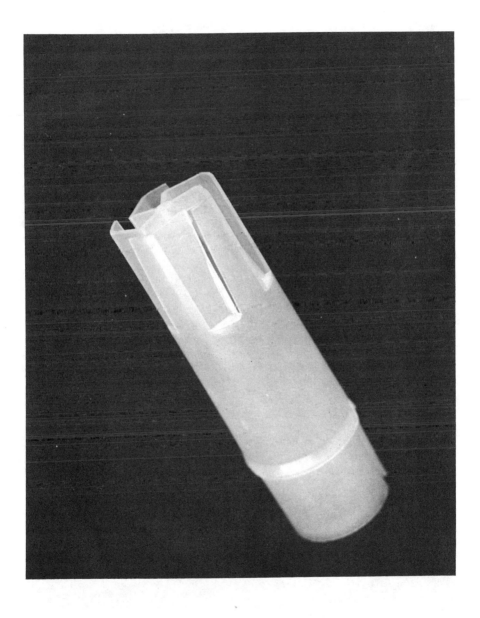

FIGURE 18: STEEL REINFORCED SPECIAL CONCRETE USED AS LINING
FOR VERTICAL TUBE EVAPORATOR AT
FREEPORT OSW TEST BED

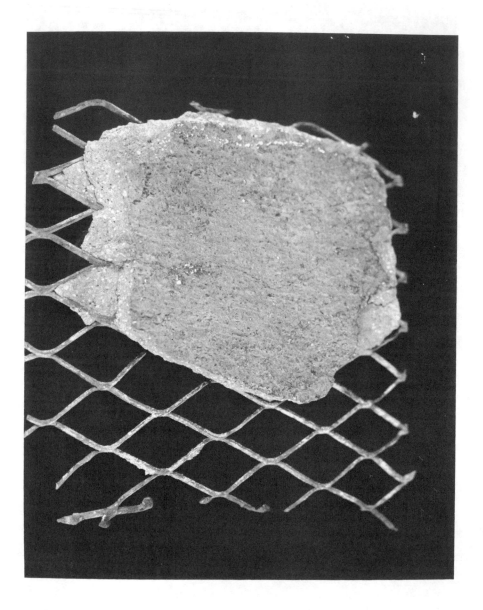

FUTURE OF NONMETALLICS IN DESALINATION

In regard to the future of these materials in desalination applications, will they be suitable and in which applications? It is obvious from the material we have covered that nonmetallics may very well have a place in intake systems, piping systems, pumps, process vessels, and possibly heat exchangers.

The corrosion resistance of these nonmetallics in general is excellent, and per dollar spent, in nearly all cases, exceeds that of metals. The big question which must be resolved and which can only be resolved by time and further testing is length of service life of these materials, and, of course, service life is a key factor in determining economics.

REFERENCES

Mallinson, John H., Chemical Plant Design With Reinforced Plastics. McGraw-Hill Book Company, Inc., New York, 1969.

Marshall, S.P., and J. Lee Brandt, "Installed Cost of Corrosion Resistant Piping." Chemical Engineering, vol. 78, no. 19 (August 23, 1971), pp. 69-82.

McGraw-Hill Book Co., Inc. (Editors of Modern Plastics Encyclopedia), Guide to Plastics (Including Handbook of Charts). New York, 1970.

McGraw-Hill Book Co., Inc., Modern Plastics Encyclopedia — 1971-1972. v. 48, no. 10-A, New York, October, 1971.

National Association of Corrosion Engineers, (Numerous Articles on Plastics by Various Authors), Materials Protection and Performance. Houston, 1960 to 1972.

National Association of Corrosion Engineers, Problem Solving with Plastics. Houston, 1971.

Plastics Pipe Institute (Literature and Standards), Society of the Plastics Industry, 9918 Sutherland Road, Silver Spring, Maryland 20901.

Society of the Plastics Industry, "Filament-Wound FRP Pipe Light and Heavy Grades A/C 1002," Proposed Standard, December, 1966.

Society of the Plastics Industry, "Filament-Wound FRP Tanks, A/C 1003," Proposed Standard, November, 1966.

U.S. Department of Commerce, National Bureau of Standards, "Custom Contact Molded Reinforced Polyester Chemical Resistant Process Equipment," Voluntary Product Standard PS15-69, November 15, 1969.

U.S. Department of Interior, Office of Saline Water, Saline Water Conversion Report (Annual Issues).

COMPARATIVE EVALUATION OF ALL MATERIALS IN DESALINATION OPERATIONS

Francis H. Coley
Office of Saline Water
U.S. Department of the Interior

ABSTRACT

The Materials Program of the Office of Saline Water includes a program for evaluation of metallic and nonmetallic materials in an environment representative of distillation plants. Much of the work is performed in a Materials Test Center at Freeport, Texas. At this site, research is conducted into corrosion mechanisms, and testing is carried out under simulated desalting conditions. A substantial portion of the test program is jointly sponsored by the OSW and industry associations. As a result of these activities, much information is being generated to show the role that many materials may assume in desalting plants. These studies are expected to continue until all major types of materials have been investigated.

INTRODUCTION

As desalination began to attract attention as an alternate source of potable water, it was only natural that processes based on evaporation should gain acceptance. The process has a natural origin that has been understood for hundreds of years. It is simple in principle and has been practiced in some fashion for many years. Its efficiency, in terms of energy requirements, leaves something to be desired; but the general attitude toward the process can be summarized as follows:

(1)   It works and can be applied in some form without extensive development programs.

(2)   It produces a product of exceptional quality — usually a very few parts per millicn of dissolved salts.

(3)   It can be used with almost any water source and, even at this date, is the only process suited to large scale conversion of seawater.

It is unfortunate that a process so useful also involves some serious problems. One of these is the problem of materials of construction; an environment of hot, saline water is one of the most aggressive corrosive media known, and a coastal environment is also known to be destructive to the exterior of plants.

This problem has a very strong influence on the cost of the water produced in terms of both initial investment and operating expense. It is so important that the process of specifying materials for a plant will eventually take into account the quality and sophistication of the personnel available to operate the plant and the quality of the services available for maintenance of the equipment and instruments used in the specific plant.

In 1964, after the Saline Water Conversion Program had existed for about twelve years, it was realized that a program of research in materials of construction offered great potential for improving the economic aspects of desalination. Accordingly the Office of Saline Water established a unit to pursue a program of research in materials. From this has grown the present Materials Program which has provided most of the existing quantitative data on behavior of materials in hot seawater.

THE INITIAL EFFORT — RESEARCH

The plans formulated for the first research effort concerned with materials took into account the fact that a major item in the initial plant cost involves tubing for heat exchangers. Since these heat exchangers are exposed to the aggressive seawater, it was felt that a program should begin with consideration of alloys for use in heat exchangers. Accordingly, it was decided to emphasize these materials in an effort combining studies of fundamental corrosion phenomena with more applied studies of metallic materials in hot seawater. As will be shown, the latter part of this activity has grown as its importance became apparent.

The first project (1) to study materials in hot seawater was undertaken with an overly optimistic point of view on the part of the OSW. A loop was constructed for the exposure of metallic coupon specimens to circulating seawater under closely controlled conditions. The term "loop" is used here, and in much of the pertinent literature, to describe a test apparatus in which corrosion test specimens are exposed to a circulating corrosive medium.

The initial phase of this operation was carried out at Harbor Island, North Carolina, and it revealed the futility of attempting such a corrosion study with inadequate controls, services and personnel. As a result of the lessons learned during that time, the unit was transferred to Freeport, Texas. There, under the supervision of personnel of the Dow Chemical Company, it was set up on a site adjoining an experimental distillation plant belonging to the OSW. The research program was then resumed for the purpose of exploring the effects of critical environmental variables such as pH, temperature, dissolved gases, and velocity on the several alloy families that might be considered for use in tubing. A further benefit was realized from the initial attempt to operate the loop in that several other problems became apparent so that studies could be directed toward solutions.

EXPANSION OF THE FREEPORT FACILITY

Shortly after the seawater corrosion program began at Freeport, Texas, three industrial associations became interested in sponsoring corrosion studies to determine the roles of their products in desalination. As a result of much planning and negotiation, it was decided to expand the small corrosion research operation at Freeport into a Materials Test Center. This was accomplished, and an important function was added with the installation of three test units to serve, respectively, the Aluminum Association, the Copper Development Association, Inc., and the Committee of Stainless Steel Producers of the American Iron and Steel Institute.

Additional test units have been added by the Office of Saline Water to provide a fairly complete capability for study of the behavior of materials in environments typical of distillation plants. The Test Center, shown in Figure 1, now includes the following:

> Facilities for treating seawater to provide feed having closely controlled pH and dissolved gas content. This part of the facility, recognized as being of paramount importance to the entire program, is equipped with elaborate controls and could serve as a model for commercial units.

> A softening system capable of removing scale-forming constituents from the seawater feed. This unit makes possible the study of materials at temperatures ranging up to 325°F. without danger of calcium sulfate scale formation.

> The original research unit, now expanded into two loops, for detailed study of the influence of environmental factors on corrosion rates.

> The three test units cooperatively sponsored by OSW and three industrial associations.

> A Metallic Materials Test Unit in which are studied such things as the effects of inhibitors and other additives on corrosion, effects of impingement on materials, effects of heat transfer, corrosion rates

in an environment of pure condensing vapor, and corrosion rates in a zone subjected to splashing brine and condensing vapor.

A Nonmetallic Materials Test Unit to study the behavior of non-metallic materials (both organic and inorganic) in hot seawater. Tests performed in this unit will involve structural plastics and composites, protective coatings, and concrete-type products.

Chemical, metallurgical and photographic laboratories for routine support of the Center and the adjacent test plant, and for the failure analyses that may be needed by the Office of Saline Water.

In all of the above work, great emphasis is placed on the need to define the environment in which the studies take place and on the need to monitor all phenomena that may affect the corrosion rate.

FIGURE 1: REAR VIEW OF OSW'S MATERIALS TEST CENTER AT FREEPORT, TEXAS

The entrance shown is at the rear of the control room,
and test units are shown on both sides of the building.

The operation of the Materials Test Center is now recognized as vital to the desalting effort. The information generated at the Center is constantly increasing, and it is obvious that the Center will have an important influence in the design and operation of plants in the future. Results obtained at the Center are published in the form of OSW Research and Development Reports (2),(3), and these are recommended to those who wish to understand the many factors affecting the behavior of the materials in desalting plants.

RELATED WORK

The research and evaluation studies in progress at Freeport are augmented by a number of additional programs. Among these are basic studies of a number of environmental, compositional and metallurgical factors that affect the behavior of materials of construction. These are sponsored in the belief that they are essential in providing the ultimate explanation for behavior that will ultimately be observed.

Another type of investigation has recently been added to those already sponsored by the Materials Division. This involves active participation by OSW personnel in a program of inspection of operating plants to identify problems within the plants and to assign priorities for study of the causes.

A system also exists for gathering, filing and retrieving information on materials behavior and failures in desalting plants. This is coupled to a service that publishes abstracts of literature having relevance to the materials-related problems of desalting. Both the information and bibliographic services are available to anyone having a legitimate interest in this field.

While these supporting programs are not as precisely oriented toward specific phenomena as those previously described, they are not lacking in importance. They provide practical insight and aid in shaping program emphasis. The prevailing attitudes toward materials problems owe much to this group.

CURRENT THINKING ON MATERIALS

As a result of the research and testing programs described in the preceding sections, some rather definite ideas on material selection are now beginning to crystallize. The following sections summarize these ideas.

The Copper Alloys

Advantages:

   (1)   The alloys are accepted in the desalting field and are used almost universally for tubing and for corrosion resistant linings in some parts of the desalting plants.

(2)　Thermal conductivity is relatively high.

(3)　These alloys display a pattern of uniform corrosion with little tendency to pit. For this reason, the lifetime of components is more predictable.

(4)　Fabricability is excellent with respect to welding and rolling.

(5)　The alloys are not sensitive to the presence of other metallic ions in seawater.

(6)　The alloys show little susceptibility to crevice and stress corrosion. These phenomena have not been problems in plants using these alloys.

(7)　With proper process control, especially with respect to pH and dissolved oxygen, a wide cost range of alloys show almost identically excellent performance.

Disadvantages.

(1)　Process control is of critical importance in determining the performance of these alloys. Dissolved gases such as oxygen, carbon dioxide, hydrogen sulfide and ammonia should be reduced to very low levels. It is also necessary to control the pH of the process stream.

(2)　Recent trends in copper prices indicate a steady increase in the cost of these alloys.

(3)　Under certain conditions, corrosion products in the plant effluents may lead to pollution problems. The concentration of these corrosion products is, of course, related to process control.

(4)　The alloys are not resistant to attack caused by turbulence or impingement. This can be a serious problem where shellfish, weeds, rocks, etc. can enter the heat rejection section of desalting plants to cause partial plugging of tubes.

Recommended Alloys:

CDA 706
CDA 715
CDA 687
CDA 194 (low dissolved oxygen
　　　　　is necessary)

The Aluminum Alloys

Advantages:

(1) These alloys offer economy in the initial investment. Studies (4) have shown that a 12 year lifetime in an aluminum alloy is comparable to an infinite lifetime for 90-10 copper-nickel.

(2) The alloys have been shown to be tolerant of a wide range of pH. Minimum pitting occurs in the range of 5.5 to 6.0 pH.

(3) There are indications that aluminum is tolerant of carbon dioxide in both liquid and vapor zones.

(4) Liquid velocity is not an important factor up to 10 feet per second so long as the velocity is maintained above a minimum of 3 feet per second.

(5) Corrosion behavior of the alloys is independent of dissolved oxygen concentration up to 1,000 parts per billion.

(6) Basically, aluminum is very versatile with a wide range of alloy compositions available.

(7) The aluminum alloys are, in general, easily worked and can be used by relatively unskilled personnel. This can be a distinct advantage for the less developed areas of the world.

(8) Due to the natural abundance of aluminum, periods of national emergency are not as likely to create serious shortages of the material.

Disadvantages:

(1) Careful engineering is necessary to avoid or to overcome the effects of galvanic coupling to more noble alloys.

(2) Galvanic relationships between aluminum alloys may cause problems.

(3) In general, aluminum alloys are considered to be sensitive to the ions of certain heavy metals. This may create engineering and process problems when it becomes necessary to remove such ions.

(4) Experience has shown that pitting of the alloys may become a problem when tubes become plugged or when deposits of scale or debris accumulate.

Recommended Alloys: (These selections are tentative; care should be taken to obtain the most recent data for evaluation and to fit the alloy to the intended application.)

3003	5005	6061
3004	5052	6163
	5454	

Stainless Steels

Advantages:

(1)   The corrosion behavior of the stainless steel alloys is independent of temperature.

(2)   Alloy behavior is, in general, not dependent upon velocity so long as a certain minimum is exceeded.

(3)   Certain pollutants, such as hydrogen sulfide and ammonia, do not produce significant attack on these alloys.

(4)   The alloys are not significantly affected by carbon dioxide in liquid and vapor phases.

(5)   The alloys exhibit a considerable immunity to "fouling." This is significant in that the overall heat transfer coefficient can be expected to remain almost constant with time.

(6)   The alloys have "eye appeal" and present a pleasing appearance.

(7)   If the circumstances that lead to pitting or cracking are avoided, the low overall corrosion rates offer the probability of almost infinite service life.

(8)   The basic ingredient of the alloys — iron — is both abundant and low in cost.

Disadvantages:

(1)   The initial cost of stainless steel is high relative to mild steel and aluminum alloys.

(2)   Installations using stainless steels must be designed to avoid crevice corrosion and pitting. This involves care in preventing stagnation of the brine, deposition of scale, and accumulation of debris.

Recommended Alloys: (These recommendations are tentative; the rapid pace of evaluation work with stainless steels makes it desirable to seek the most recent data on their behavior in this environment.)

216   329
316   E-Brite 26-1
Alloy 20

Alloys to Avoid:

409
410
430

Titanium

There is little information available on the behavior of titanium alloys in desalting environments. Such tests as have been made have not subjected the material to the extreme conditions that a premium material should be expected to withstand. However, some statements can be made.

Advantages:

(1)   Titanium alloys are known to be inherently very resistant to corrosive attack. In situations when the attack is uniform the titanium may last indefinitely. Dissolved gases can probably be ignored, and close control of pH will not be needed.

(2)   Titanium is very strong.

(3)   Pollution problems should be minimized through the use of titanium, since the material should not contribute appreciable concentrations of harmful corrosion products to the plant effluent.

(4)   Titanium alloys will be relatively immune to corrosive attack by the pollutants that might occur in the seawater feed.

(5)   Cleaning operations to remove scale from heat transfer surfaces could be conducted without expensive precautions to protect titanium surfaces.

(6)   Titanium alloys are resistant to attack under turbulent or impingement conditions.

Disadvantages:

(1)   The alloys are relatively expensive.

(2)   Titanium alloys have a relatively low thermal conductivity.

(3)   Under certain conditions the titanium alloys may undergo corrosion by crevice and/or pitting attack. Careful design will be necessary to avoid such problems.

(4)   The titanium alloys are somewhat difficult to fabricate. This may require special training for the workers engaged in assembly of the plant.

Predictions: Titanium alloys will find application as tubing in the heat rejection section of plants where high concentrations of dissolved oxygen and pollutants may create problems with copper alloys. When partial plugging by shells, rocks, etc. is likely, the material should also be less susceptible to penetration in the turbulent areas. It seems obvious that titanium will be competitive with stainless steel and the more expensive copper alloys.

Carbon Steel

A few years ago it was felt that an economical desalting operation required all vessels to be built of carbon steel and that no protection could be given to the material. Reports from operating plants have shown, however, that such treatment leads to material failure and to excessive maintenance expenditures. Carbon steel has now become the greatest single problem area in distillation plants.

Advantages:

(1)   Carbon steel is the least expensive structural alloy in terms of initial cost.

(2)   The material is abundant, and most welders and fabricators are experienced in its use.

(3)   A certain amount of evidence indicates that carbon steel can be used without protection in plants operating at low temperature (to $190°F$.) and using additives for scale control.

Disadvantages:

(1)   There is a paucity of positive evidence indicating that carbon steel can survive without protection in a plant using acid treatment for scale control.

(2)   Rigid control of the environment within the plant is essential. Carbon steel is very sensitive to pH, velocity and dissolved oxygen. In addition, it is felt that poor decarbonation also contributes greatly to corrosion of carbon steel in areas above the liquid line in a plant. It is also suspected that copper salts (corrosion

products) may produce severe localized attack and that
bacteria may even be a factor in the deterioration of the
material.

Low Alloy Steels

Research has shown (5) that some low alloy structural steels of the self-painting
type may be more satisfactory than bare carbon steel in resisting corrosive attack. The superiority of these steels is lost, however, unless the dissolved oxygen content of the feed water is maintained at a very low level.

Recent information (6) has also shown that some low alloy steels are of interest
for tubing application in Japan. Such alloys should accomplish much toward
reducing plant investment, but it must be understood that much work remains
to be done before these alloys are qualified for plant use.

NONMETALLIC MATERIALS

The role of nonmetallic materials in desalination is not well established, although
there is a widespread belief that they will prove useful. Research is in progress,
but the process of fitting specific formulations to specific applications is only
beginning. Thus, one can only mention the applications that are foreseen. Both
organic and inorganic materials are under consideration.

Organic Materials

Many possibilities for use of organic polymers exist. These applications may be
in the form of structural components or as coatings.

Structural Applications: It is believed that composite materials, such as fiber
reinforced plastics, are very promising candidates to replace metallic materials
in such components as piping, tube sheets, baffles, panels, product troughs,
weirs, etc. Such materials offer high resistance to corrosive attack, freedom
from maintenance, and relatively low cost. It is likely, however, that tempera-
ture-strength relationships will limit the extent to which these materials can be
applied in plants. Needless to say, the ultimate selection will require an economic analysis to provide the optimum choice.

Another "structural" application may involve the use of either concrete impregnated with a polymer (PIC) or a polymer concrete (PC) in which a polymer becomes the binder for an inorganic aggregate. The use of these materials
for vessels, pipes and panels is envisioned. These materials seem to offer strength
and resistance to brine much superior to that of ordinary concrete, but it is not
yet clear that these materials are economically competitive with the more common materials or that the polymers will survive for sufficient time to make the
application worthwhile.

In connection with these applications, it should be remembered that mere substitution in which a metallic component is replaced by a nonmetal may not represent a good application. Ideally, these materials should be included in the original design so as to take maximum advantage of their unique properties.

Organic Coatings: The most frequently considered application for coatings of this type is for the protection of carbon steel in the vapor zones or, at lower temperatures, within the splash zones of distillation plants. While a wide variety of formulations are available and under test, it is too early to determine what the ultimate coating will be. The following points will be of importance in selection of a coating, however.

Can the coating be applied and repaired in the field? Coatings applied in a factory should be superior in quality to those applied in the field, since a factory can be expected to provide better quality control. However, since no coating can be expected to last forever, it is most desirable that it be suitable for field repair and application.

What surface preparation is necessary? Surface preparation may range from rough brushing of surfaces to very careful sandblasting to provide a bright finish. Thus, surface preparation can have a significant effect in the cost of a coating, and money spent to assure good surface treatment is usually well invested.

How difficult is application? The method of application may involve brushing, rolling, or spraying. However, the most important consideration during application is probably that of getting a uniform coat of the optimum thickness. This makes it mandatory that rigorous inspections be carried out by qualified technicians.

What cure is needed? Some coatings will be cured by the evaporation of solvent, some may require heat, and others may even require exposure to some type of radiation. In specifying coatings, the designer should be aware of the curing requirements. He should also realize that safety is a factor in this process, since fumes may present a hazard to the health of workers because of toxicity or the possibility of explosion.

Dropwise Condensation Promotion: Certain organic materials have the ability to cause metallic surfaces to become hydrophobic rather than hydrophilic. This alters the condensation process to a dropwise form instead of a filmwise mode. This results in much higher rates of heat transfer and can produce important increases in plant output. The utility of this idea is still to be determined, however, since there remain important questions as to durability and cost of such coatings.

Miscellaneous Applications: It is likely that organic materials will be used in many other ways. For example, it has been shown (7) that evaporation ponds used for brine disposal in some locations may require sealants to prevent seepage into either the adjacent soil or the formations below. Plastic in sheet form can be used for this service, and it also may be possible to impregnate a thin layer of soil to create an impervious layer. Such materials might also be applied as linings in ditches used to convey a brine to a disposal site.

Inorganic Nonmetals

The inorganic nonmetallic materials of interest to the desalination industry are members of the family of materials designated as mortars or concretes. A rather wide variety of concretes exist to provide properties suitable for a number of different applications, and it appears that these interesting materials will be used as protective coatings and as structural components.

Protective Coatings: Products having a high content of monocalcium aluminate or aluminum silicate are now being used as lining materials to protect carbon steel. In portions of the plant, a coating of this material may be applied to a thickness of about 1 inch on all of the carbon steel surfaces below the entrainment separators. Field experience to date indicates that these materials, properly applied, offer excellent protection to carbon steel. The coatings are costly, however, and studies must continue to establish the maximum temperature at which they will prove useful.

Structural Applications: As yet there has been no successful application of concrete for vessel construction, but efforts in this direction have actually been minimal. It somehow seems that interest is growing in developing designs suited to the use of concretes. At the same time, work is in progress to determine just what type of concrete should be used. This basic material might be either a high grade concrete used without protection or a low cost material used with a protective coat or impregnant. The ultimate choice will be based on the outcome of experiments now in progress and the economic analysis that must follow.

Possible Problems: In the course of operating a desalting plant occasional cleaning of the metallic surfaces may be required to maintain heat transfer coefficients at a practical level. Unfortunately, the solutions used in these cleaning operations are likely to have an adverse effect on concretes and mortars. Therefore, it is conceivable that there will always be a need to impregnate or coat the concrete products to protect them during the cleaning operation.

WHAT THE FUTURE HOLDS

One of the goals of those involved in the study of materials for desalting plants is to make possible the selection of the optimum combination of materials for a given plant. This goal has not been achieved, but it does appear certain that it can be reached. In the meantime, there is a growing insistence that plants be made more reliable. This will require that the old concept of "minimum plant investment" be abandoned for the present since the plants built by this standard have not performed as well as had been expected.

The studies that have been made to date have contributed important insights into the materials problems in desalting. Even though all the answers are not available, it is now possible to predict the following:

(1)   The use of bare carbon steel will be abandoned until such time as rigorous process control procedures can be followed.

(2)   Metallic materials other than copper alloys will find wider use in plants. Titanium will undoubtedly find fairly rapid acceptance for some tubing applications. Aluminum alloys will find wider acceptance, especially at temperatures below $200°F.$; this application may include both vessel construction and tubing. Stainless steels will find application in portions of the plants, but their acceptance may be delayed until some of the better alloys can be produced at lower prices. Pollution considerations may be important in diminishing the use of copper alloys.

(3)   Greater emphasis will be placed on the use of nonmetallic materials, and more effective candidates will be found. Composites will prove especially interesting.

(4)   There will be an increase in emphasis on process control and on the use of superior operating personnel in locations where services and personnel of high quality are available. The philosophy of operation will more closely approach that of a chemical plant operated for profit. In areas where sources and personnel are not of high quality, a "foolproof" design will be used. In other words, future purchasers of plants can be expected to choose either quality operation or materials that will withstand poor operation.

(5)   Attitudes toward the cost of desalting will become more realistic. The insistence on "low cost" will change to insistence on reliability as the time cost of operating and maintaining the early type of "minimum investment" plant becomes known.

REFERENCES

(1)   Battelle Memorial Institute: "Investigation of Corrosion in Hot Sea Water in an Experimental Loop Apparatus," Office of Saline Water Research and Development Program Report No. 225 (December 1966).

(2)   Dow Chemical Company: "Seawater Corrosion Test Program," Office of Saline Water Research and Development Program Report No. 417 (March 1969).

(3) Dow Chemical Company: "Seawater Corrosion Test Program — Second
     Report," Office of Saline Water Research and Development Progress
     Report No. 623 (December 1970).

(4) E.T. Wanderer and M.W. Wei, Metals Engineering Quarterly, Vol. 7, No. 3
     (1967), p. 30.

(5) Dow Chemical Company: Unpublished data to appear in reports on con-
     tinuation of Seawater Corrosion Test Program as described in References
     (2) and (3).

(6) A. Takamura, Kobe Steel Ltd., Private Communication.

(7) Bureau of Reclamation, United States Department of the Interior: "Pond
     Linings for Desalting Plant Effluents," Office of Saline Water Research
     and Development Program Report No. 602.

OVERALL ECONOMIC CONSIDERATIONS
OF DESALINATION

J.J. Strobel
Office of Saline Water
U.S. Department of the Interior

ABSTRACT

The economics of large scale desalination plants are presented and the estimates of producers' costs for several size plants are given.

INTRODUCTION

First, I would like to tell you about the Office of Saline Water and what its mission is, because I believe this will lay the basis of sources of information which are drawn on in this paper. OSW's major mission is to develop low-cost means of obtaining fresh water from saline or salt water sources; to develop processes to obtain this water on a large-scale basis for municipal, industrial, agricultural, and other beneficial uses.

Since the early 1950's the Office has conducted an R&D program with that the major objective. This program is conducted primarily through contracts and grants with universities, industrial companies, research institutes, and related organizations. As the R&D progressed, the need became apparent to compare costs between processes and to project costs into the future. Since ways of getting low-cost desalting are under study, present and future water costs both for desalting and for conventional water supply become extremely important.

This paper is based on a number of sources of desalting economics information such as the R&D program, including pilot plant and experimental plant data, data from actual plants, including demonstration plants, and from design and cost studies on desalting plants for the future. A number of preliminary feasibility studies of desalting applications have been supported by OSW, as well as studies of current and projected process economics.

The accompanying figures illustrate the relationships between many of the more significant plant parameters. To assist the reader in further investigation, a list of the references is included. I would like to mention the types of OSW contractors for the economics studies. There are firms such as A&E firms, consultant firms, research institutes, and some government organizations. Examples are Burns and Roe, Bechtel, Fluor, Catalytic, Kaiser, Ralph M. Parsons, Hittman Associates, A.D. Little, Stanford-RI, Southwest Research Institute, and Oak Ridge National Laboratory.

OVERVIEW

In this paper, I shall address myself to descriptions of the major elements of desalting operations costs and the general procedures for determination of their amounts and variations. My aim shall be to explore the more significant relationships between plant design, site, energy form, manpower, and costs, and how these factors combine to determine the cost of the final product — desalted water.

Although I shall, for the most part, discuss distillation and membrane plants, the concepts and methods I shall present are applicable to all desalting operations including freezing, ion exchange, and other processes. In addition, the methodology may be applied to both inland brackish desalting plants as well as coastal seawater desalting installations.

Since the purpose of economic considerations of a given desalination operation frequently is economic comparison with some alternative operation, it becomes necessary to establish various ground rules concerning the manner in which different operations will be treated. Table 1 lists these major ground rules.

TABLE 1: MAJOR DESALTING PLANT ECONOMIC AND DESIGN PARAMETERS

	Units
Economic parameters	
Interest rate	%
Insurance and tax rate	%
Electrical power cost	mills/kwh
Thermal energy cost	$¢/10^6$ Btu
Labor costs	$/hour
Chemical costs	$/pound
Design parameters	
Plant design lifetime	year

(continued)

TABLE 1: (continued)

	Units
Plant load factor	%
Feedwater temperature; salinity	°F.; ppm
Reject brine concentration; product to feed or recovery ratio	ppm; fraction
Required product salinity	ppm
Energy utilization	lb. product/1,000 Btu; kwh/1,000 gal. product
Tubing; membrane performance criteria	Btu/hr./ft.^{2} gal./day/ft.^{2}
Tubing; membrane lifetime	year

FIGURE 1: WORLD DESALTING MARKET AS OF JANUARY 1, 1971

FIGURE 2: PRINCIPAL AREAS BEST SERVED BY DESALTING

With the ground rules established, specific plant designs may be engineered and evaluated. The costs associated with these designs may be grouped most conveniently into two major cost centers. These are:

(1)   Fixed costs
(2)   Operating and maintenance costs

I will elaborate more fully on the nature of these elements of cost, their determination and degree of significance, and contribution to the overall cost of desalted water; but first let us review some of the existing and projected desalting plant installations.

Why do we consider desalting economics so important? A major reason is that if through the R&D efforts the low cost goals can be reached, and if along with that desalting technology becomes better known and used, there will be a very great increase in the amount of desalting for water supply. Figures 1 and 2 and Table 2 describe current and projected uses of desalting technology. To get some perspective on projected uses, we need to know something about what the status of desalting is now.

TABLE 2: PROJECTED ROLE OF DESALTING FOR U.S. WATER SUPPLY

NEW DESALTING AS CAPACITY (MGD) AND AS PERCENT OF TOTAL AUGMENTATION

REGION	1980 CAPACITY (MGD)	%	2000 CAPACITY (MGD)	%	2020 CAPACITY (MGD)	%	CUMULATIVE DESALTING
SOUTH ATLANTIC GULF	20	4	380	6	800	9	1200
TEXAS GULF	60	4	415	9	975	15	1450
RIO GRANDE	40	8	135	13	225	18	400
LOWER COLORADO	150	7	200	13	300	17	650
CALIFORNIA	160	9	1490	19	2150	24	3800
OTHER REGIONS	80	16	470	4	1250	3	1800
PROJECTED TOTAL DESALTING	510		3090		5700		9300
NATIONAL AUGMENTATION	5,000		33,000		69,000		107,000

1971 DESALTING – 56 MGD

DESALTING PLANT INSTALLATIONS — EXISTING AND POTENTIAL

Currently, there are over 750 desalting plants installed or under construction throughout the world; Figure 1 shows these locations. Of these, 686 are distillation plants, while the balance are primarily membrane plants with the electrodialysis process predominating. The total world production of these plants is about 313 mgd.

The distillation plants range in size from 25,000 gpd to 30 mgd, while the ED plants range from 26,000 gpd to 1.2 mgd. Figure 1 shows where the bulk of the capacity is located.

Plants reported as sold in the 1969 to 1970 period include 49 distillation plants having capacities ranging from 27,000 gpd to 30 mgd. Seventeen membrane plants were sold with capacities from 29,000 gpd to 1.2 mgd. In addition to these, nine vapor compression type desalting plants and two reverse osmosis plants were sold.

The cumulative projected potential for desalting in the United States is illustrated in Table 2. Note that the current capacity is only a ninth of that projected for 1980. These estimates of future requirements are based on a detailed OSW planning model which simulates the requirements for desalting in various hydrologic regions. The predominant feedwater for these plants in terms of daily capacity is seawater. In later years, desalted recycle water is expected to contribute greatly to the supply of fresh water. The aggregate capacity of desalting plants using saline groundwater feed is expected to be about 400 mgd in the year 2020.

In general, these plants will increase in size with time, and we may expect that the bulk of capacity will be in plants greater than 50 mgd each.

Based on an analysis of near-term U.S. water supply requirements, both for maintaining quality due to deteriorating supplies and requirements for new water, we may expect a number of areas of the U.S. to be candidates for desalting plants; these are shown in Figure 2. Although a number of the plants will be membrane plants operating with brackish water feeds, the bulk of capacity will be made up of distillation plants operating on seawater feed.

These desalting plants are expected to produce water within the cost ranges illustrated in Figure 3. The ranges shown take into account uncertainties in the costs of energy and capital, the plant location, and the degree of technological advancement attained over the period examined. Average estimated costs for desalted water are expected to be 60¢/kgal. in 1980 based on late 1971 dollars.

ECONOMIC FACTORS IN DESALTING

The cost of desalted water is strongly influenced by the economies achieved with increasing size. This is apparent in Figure 4 which depicts relative water costs for various sizes of plants and two major processes. The predominant impact of increasing the plant size is to lower the unit charge for labor and capital. The costs shown in Figure 4 are based on near-term fuel prices and fixed costs predicated on an interest rate of 5⅜% and a plant life of 30 years. For the distillation process, a 50 mgd plant was used for the reference size. A 10 mgd plant was used as reference for the membrane plants comparison. The reference cost for the 50 mgd plant is 69¢/kgal., the reference for the 10 mgd membrane plant is 28¢/kgal.

FIGURE 3: COSTS OF WATER SUPPLIES

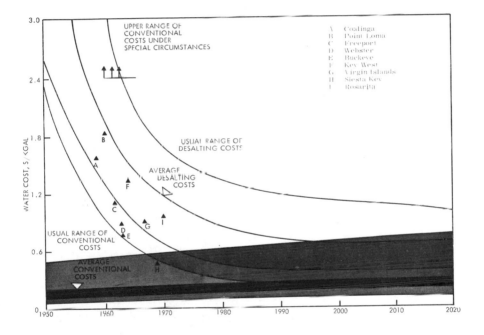

METHODOLOGY

A convenient format for estimating the cost of desalted water is illustrated in Figures 5 and 6. This format utilizes major cost centers — fixed costs and operation and maintenance costs; these cost centers are then further subdivided. Note that capital costs are subdivided into depreciable items and nondepreciable items. The latter are assumed not to lose value over the life of the plant. Recurring costs include insurance and taxes which must be included as a portion of the cost of owning and operating the plant. Note that "replaceable items" are treated as part of capital cost; such items include allowances for

FIGURE 4: INFLUENCE OF PLANT SIZE ON WATER COST

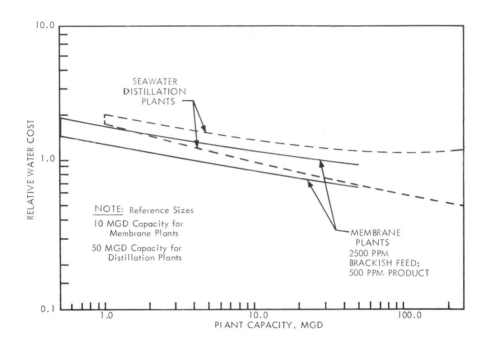

major items of equipment. A detailed discussion of this cost estimating procedure is contained in the OSW R&D Report No. 264.

The cost of desalted water also may be illustrated by division into three major
categories; these are capital plus recurring costs, energy costs, and labor plus
maintenance costs. Figure 7 shows how the relationship between these costs
change as the size of the plant is varied. Note that for the larger plant, the
energy costs form a significantly higher fraction of the total costs. Naturally,
these allocations are quite sensitive to the cost of energy, the interest rate, and
the plant life and load factor.

Figure 8 demonstrates the relationship between fuel cost and its contribution
to cost increases, but not at a linear rate. Since the plant designer has the op-
tion of increasing the performance ratio of the plant in order to decrease the
fuel consumption, he can trade off fixed charges for capital against the variable
cost of energy. In this manner, specific optimized designs having the most economical overall costs may be derived. Note that the more costly, higher performance ratio plant has the lower fractional energy cost since its fixed charges
are proportionately higher as is its efficiency. Several design and cost optimiz-
ing computer programs are available currently.

FIGURE 5: WATER COST SUMMARY

SEAWATER DISTILLATION (MSF) WITH STEAM BOILER						
PERFORMANCE RATIO = 12.3				Water Pl: 1 MGD 65 F / Capacity Factor: 90% / Output: 3.3 x 10⁵ Kgal/yr / Process: Seawater Distillation		
INTEREST RATE = 5-3/8%; PLANT LIFE = 30 yr						

		CAPITAL COSTS in $10³	CARRYING CHARGE MULTIPLIER	ANNUAL COSTS in $10³	WATER COST	
					¢/Kgal	%
CAPITAL COST CENTERS						
	1. Desalting equipment	1280.	6.787	86.9	26.3	
	2. Replaceable items	----	*			
	3. Brine disposal	----				
	4. Water treatment	53.5	6.787	3.6	1.1	
	5. Intake and discharge	----				
	6. Steam supply	252.	6.787	17.1	5.2	
	7. Power Supply	----				
	8. Structures and improvements	32.2	6.787	2.2	0.7	
	9. Indirect capital costs	299.1	6.787	20.3	6.1	
	10. Land costs	150.	5.375	10.2	3.1	
	11. Working capital	48.8	5.375	3.3	1.0	
	TOTAL CAPITAL COSTS	2115.6				

			PERCENT			
RECURRING COST CENTERS						
	12. Taxes		----			
	13. Ordinary insurance		0.25	4.8	1.5	
	14. Nuclear insurance		----			
	TOTAL ANNUAL FIXED COSTS			148.4	45.0	35.6

OPERATION AND MAINTENANCE COST CENTERS						
	15. Operating labor (incl. fringe benefits @15%) ⎫			75.2	22.8	
	16. Maintenance labor (incl. fringe benefits @15%) ⎬					
	17. General and administrative			19.6	5.9	
	18. Supplies and maintenance materials			10.5	3.2	
	19. Chemicals			13.6	4.1	
	20. Fuel (@ 50¢/MM Btu)			138.0	41.8	
	21. Steam			----	----	
	22. Electric power (@ 7 mills/Kw-hr)			11.6	3.5	
	23. Other			----	----	
	TOTAL OPERATION AND MAINTENANCE COSTS			268.5	81.3	64.4
	TOTAL FIXED PLUS O&M COSTS			416.9	126.3	100.0

*Carrying charge multiplier = Interest + Sinking Fund Factor @ i = 5-3/8% for 30 yr
Carrying charge multiplier = 5.375 + 1.412 = 6.787%

FIGURE 6: WATER COST SUMMARY

SEAWATER DISTILLATION (MSF) WITH STEAM BOILER					
PERFORMANCE RATIO = 12.3					
INTEREST RATE = 5-3/8%; PLANT LIFE = 30 yr					

Water Pl: __10__ MGD __65__ F
Capacity Factor: __90%__ .
Output: __3.3 x 10⁶__ Kgal/yr
Process: Seawater Distillation

	CAPITAL COSTS in $10³	CARRYING CHARGE MULTIPLIER	ANNUAL COSTS in $10³	WATER COST c/Kgal	WATER COST %
CAPITAL COST CENTERS					
1. Desalting equipment	9731.0	6.787	660.4	20.0	
2. Replaceable items	------	*			
3. Brine disposal	------				
4. Water treatment	347.0	6.787	23.6	0.7	
5. Intake and discharge	------				
6. Steam supply	1512.0	6.787	102.6	3.1	
7. Power supply	------				
8. Structures and improvements	139.0	6.787	9.4	0.3	
9. Indirect capital costs	1978.7	6.787	134.3	4.1	
10. Land costs	500.0	5.375	33.9	1.0	
11. Working capital	344.5	5.375	23.4	0.7	
TOTAL CAPITAL COSTS	14,552.2				
RECURRING COST CENTERS		PERCENT			
12. Taxes		----			
13. Ordinary insurance		0.25	34.3	1.0	
14. Nuclear insurance		----			
TOTAL ANNUAL FIXED COSTS			1021.9	30.9	35.07
OPERATION AND MAINTENANCE COST CENTERS					
15. Operating labor (incl. fringe benefits @ 15%)			105.5	3.2	
16. Maintenance labor (incl. fringe benefits @ 15%)			45.0	1.4	
17. General and administrative			47.2	1.4	
18. Supplies and maintenance materials			58.6	1.8	
19. Chemicals			136.3	4.1	
20. Fuel (@ 50¢/MM Btu)			1380.0	41.8	
21. Steam			----	----	
22. Electric power (@ 7 mills/Kw-hr)			115.5	3.5	
23. Other			----		
TOTAL OPERATION AND MAINTENANCE COSTS			1888.1	57.2	64.93
TOTAL FIXED PLUS O&M COSTS			2910.0	88.1	100.00

Left margin labels: DEPRECIABLE (items 1–9); NON-DEP. (items 10–11)

*Carrying charge multiplier = Interest rate + sinking fund factor @ i = 5-3/8% for 30 yr

Carrying charge multiplier = 5.375 + 1.412 = 6.787%

FIGURE 7: WATER COST DISTRIBUTION

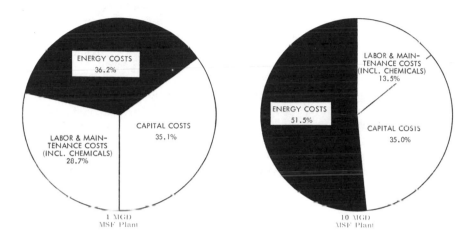

FIGURE 8: FUEL COST FRACTION OF PREDICTED PRODUCT WATER COSTS FOR SINGLE PURPOSE 10 MGD PLANTS

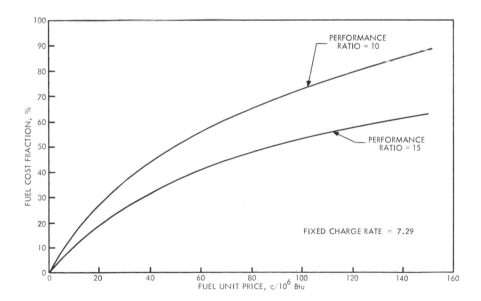

No discussion of desalted water costs would be complete without some mention of the importance of the plant load factor. Figure 9 illustrates the effect of the plant availability on the cost of water. It is important to note that, although the cost of water rises rapidly with falling availability, the total annual costs decrease. Naturally, this is due to the decrease in operating costs for fuel, electricity, chemicals and some supplies; fixed charges continue to accumulate, of course, as do charges for the operating crews unless they can be called in on a part-time basis. The plant could be put completely on standby or mothballed, in which case some additional savings in labor could be realized. Such a plant would have a low design-point availability.

For normal water supply purposes, attainment of high plant load factors is very desirable since extra plant capacity and storage capability is required for each incremental loss of plant availability. For this reason, the use of corrosion resistant materials of construction, the specification of components having great durability, and the selection of systems with a high degree of maintainability and short repair times are all beneficial with respect to realizing high plant load factors.

FIGURE 9: EFFECT OF PLANT AVAILABILITY ON COSTS FOR 10 MGD DISTILLATION PLANT

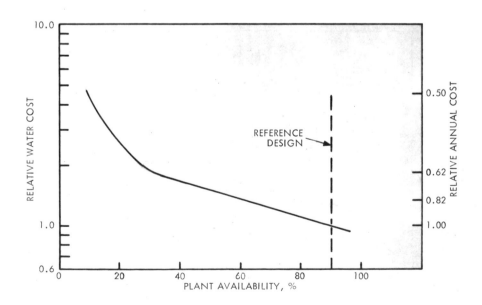

COST ALLOCATIONS IN A SEAWATER DISTILLATION PLANT

An example of the allocation of costs for an operating plant is shown in the bar chart of Figure 10. These data are based on actual costs experienced in the operation of the Guantanamo Bay (GITMO) distillation facility which consists of three separate plants having an aggregate rating of 2¼ mgd and a performance ratio near seven pounds of water per pound of steam. The facility consists of the 28 stage Point Loma multistage flash evaporator and two newer 15 stage evaporators. The Point Loma evaporator originally was operated as an OSW test unit at Point Loma, California, prior to being dismantled and shipped to Guantanamo to meet fresh water requirements. Each of these plants is rated at 750,000 gpd.

FIGURE 10: DISTILLATION PLANT'S COST ALLOCATION

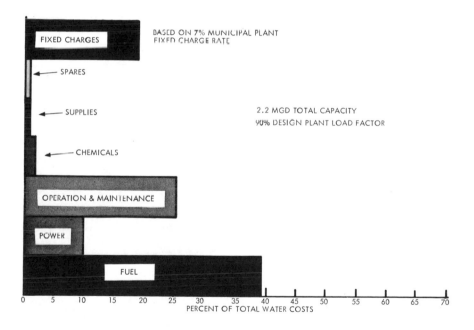

All three plants share a common seawater circulating system and all receive steam from the same source — backpressure turbine exhaust steam blended with desuperheated steam from the main boilers. In addition to producing water, this facility produces electrical power used throughout the facility. In computing the fixed charges for installation, it was assumed that the plant capital cost was that of a single 2¼ mgd distillation plant rather than the three units actually installed. This was done to reflect current design trends which dictate using the largest economical production unit. Figure 10 shows again the significance of fuel and power costs which total nearly 50% of the overall cost. In unit costs, this charge is about 83¢/kgal. for this location and type of fuel.

A significant contributor to the excellent results realized at this installation is the relatively high plant availability whose mean monthly value is estimated at 86%. This value, of course, also takes into account outages not only in the desalting plant, but also those in the steam generation and electrical power generator facilities. Another point of interest is the low percentage contribution due to chemicals, supplies and spare parts, which total approximately 4%.

The percentage of labor for operation and maintenance reflects an allocation of labor between the evaporator system itself and the steam generator-turbogenerator system. This sharing of personnel results in about one-half the staff being charged to water production and illustrates one of the benefits of a dual purpose type of plant in which both power and water are produced.

As mentioned earlier, corrosion resistant materials, durable pumps, and other components are important in realizing high plant load factors. Based on the experience at the GITMO distillation plant over the last six years, the materials shown in Table 3 appear to be sound choices for the service indicated at this location. These recommendations are conditional inasmuch as each prospective desalting site should be evaluated thoroughly with respect to anticipated operating conditions, corrosion engineering features, and the most up-to-date corrosion information.

TABLE 3: TENTATIVELY RECOMMENDED MATERIALS OF CONSTRUCTION

Service	Material
Evaporator vessel (170° to 200°F.)	Epoxy lined carbon steel
Evaporator vessel (<170°F. - deaerated)	Carbon steel
Water boxes	Stainless steel or Cu/Ni or Cu/Ni lined steel
Seawater butterfly-type valve	Bronze body, Cu/Ni disc, Buna N seat (for throttling - no Buna N)
Deaerated brine or distillate valve	Cast iron body, Al-bronze disc, Buna N seat (for throttling - no Buna N)
Air ejection condensers	Type 316 stainless steel tubing and shells
Brine heating surface	90-10 Cu/Ni or Al-brass (ASME SB111)

(continued)

TABLE 3: (continued)

Service	Material
Heat recovery surface	90-10 Cu/Ni or Al-brass
Heat rejection surface	Al-brass at ~5 ft./sec.; 90-10 Cu/Ni at 7 ft./sec. (max.)

Current Recommendations

Evaporator vessels above demisters	Epoxy-lined carbon steel
Remainder of all vessels	Concrete-lined carbon steel

INTEGRATION OF POWER AND DESALTED WATER WITH CONVENTIONAL SUPPLIES AND POWER PLANTS

You will recall that desalting plants may be operated in conjunction with power plants and, of course, in conjunction with conventional water supplies. Such combinations result in greater economy and in decreased fluctuation In water supply. Alternatively, conjunctive-type plants may be used as droughtproofing systems making water available in dry years and staying inoperative in wet years. An example of such an application is shown in Figure 11.

This system consists of a nuclear steam generator, a turbogenerator system, an evaporator plant, and reservoir system. In this particular example, the system may be operated to produce only power or it may be operated to produce both water and power in varying ratios, depending on the projected need for water to meet firm yield requirements and on the existing demand for power. This system can take advantage of the low cost of off-peak interruptible electrical power and steam which may be bleed steam or exhaust steam from a low pressure backpressure-type turbine.

The design of a conjunctive system of this type and size necessarily rests on the analysis of hydrologic records and predictions of firm yields and future demands for both water and power. The cost of water calculated for this application also depends strongly on various assumptions including fixed charge rates, useful life-times, power and steam costs, and the method of allocating costs. Various studies of such conjunctive systems which are base-loaded show that the most economical performance ratios are much lower than those for base-loaded plants with high plant factors. Essentially, this is due to the trade-off in the higher costs of capital for increased performance ratios versus the lower costs of energy for interruptible off-peak service.

Another important factor determing cost in conjunctive applications is the probability value selected for meeting the demand in any period. Higher values increase the size and cost of the plant since the user has a lesser desire to risk any shortage. A thorough discussion of cost allocation methods for dual purpose distillation plants may be found in OSW R&D Report No. 490.

FIGURE 11: CONJUNCTIVE USE

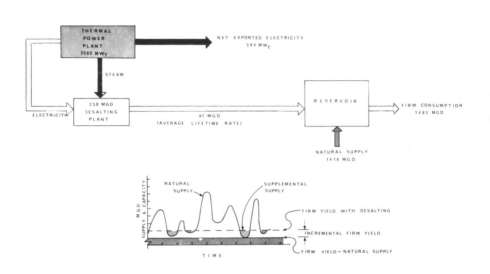

Although I have not emphasized recycled water recovery to any great degree, it should be noted that there is a growing interest in the desalting of such feed-stocks for use in both municipal and industrial applications. As shown earlier, an aggregate recycle water desalting capacity of 190 mgd by the year 2020 is anticipated. Whether these plants will be multipurpose conjunctive systems or single purpose base-loaded systems is not clearly defined at this time; however, it is expected that some large conjunctive use systems will be installed using either distillation or membrane technology.

ACKNOWLEDGEMENTS

The assistance of Mr. Robert Blevitt of Hittman Associates, Inc., and Mr. Herman Sturza of the Office of Saline Water, Department of the Interior, in assembling and preparing materials for the lecture is gratefully acknowledged.

REFERENCES

(1)   "Guidelines for Uniform Presentation of Desalting Cost Estimates," The R.M. Parsons Co., OSW R&D Report No. 264.

(2)   "Multistage Flash Distillation Desalting State-of-the-Art (1968)," Hittman Associates, Inc., OSW R&D Report No. 491.

(3)   "Vaccum Freezing Vapor Compression Desalting State-of-the-Art (1968)," Hittman Associates, Inc., OSW R&D Report No. 491.

(4)   "Cost Analysis of Six Water Desalting Processes," Stanford Research Institute, OSW R&D Report No. 495.

(5)   "Optimum Operation of Desalting Plants as a Supplemental Source of Safe Yield," Utah University, OSW R&D Report No. 528.

(6)   "Desalting Cost Calculating Procedures," Southwest Research Institute, OSW R&D Report No. 555.

(7)   "Electrodialysis Desalting State-of-the-Art (1969)," Hittman Associates, Inc., OSW R&D Report No. 610.

(8)   "Reverse Osmosis Desalting State-of the Art (1969)," Hittman Associates, Inc., OSW R&D Report No. 611.

(9)   "Survey of the Ion Exchange Process for Desalination Applications," Control Systems Research, Inc., OSW R&D Report No. 616.

(10)   "Cost of Steam Supply for Desalting Using Coal Fired Boilers," Sanderson and Porter, Inc., OSW R&D Report No. 634.

(11)   "Economic Analysis of the Membrane Water Desalting Processes," Stanford Research Institute OSW R&D Report No. 638.

(12)   "A Study of Conjunctive Operation of Nuclear Dual Purpose Desalting Plants to Serve New York City Metropolitan Region," S. Shiozawa, International Atomic Energy Panel, Vienna, Austria, April 5-8, 1971.

ECONOMICS OF MIDDLE-SIZE DESALTING PLANTS

William J. Schwarz, Jr.
Universal Desalting Corporation

ABSTRACT

The economics of distillation mode seawater desalting plants in the capacity range of 50,000 gallons per day to 1 million gallons per day are presented. Factors to be considered in the design of these facilities are reviewed; special emphasis is placed on Caribbean conditions. Water cost projections and sample calculations are offered for two plant capacities: 100,000 gpd and 50,000 gpd.

INTRODUCTION

To outline the scope of the subject, this discussion of the economics of middle-size desalting plants may well begin by defining the terms "economics" and "middle-size."

Economics, according to the dictionary, is "the science that deals with the production, distribution and consumption of wealth and with various related problems of labor, finance, and taxation." As applied to this discussion of desalting, economics embraces the factors which influence the cost of pure water as produced by distillation methods. Specifically, we will evaluate factors which influence the costs of desalting by distillation and then calculate practical examples to answer the usual question, "How much does it cost?" Answers, again to be specific, will be computed in dollars per thousand gallons of product water.

Middle-size also demands definition. "Middle," again returning to the dictionary, is defined as "equally distant from the extremes: forming a mean." The mid-range of limits reviewed here extends from 50,000 gpd to 1 mgd. Persons

168

who have had an opportunity to work at the operating level in desalting are aware of the implications of size.

Factors that determine the cost of producing pure water by distillation in all size ranges include both fixed and operating costs; the components of these costs are listed in Table 1. Costs are usually calculated on a yearly basis, although monthly, quarterly or other periods could be used.

TABLE 1: COST FACTORS

Fixed Costs
Capital costs
Taxes
Insurance

Operating Costs
Fuel costs
Power costs
Operating labor costs
Chemical costs
Maintenance costs
Plant availability

FIXED COSTS

Fixed charges that must be included in the total cost of producing water from a desalting plant are predominantly the result of capital costs, consisting mainly of amortization and interest to recover the installed cost of the plant. The amount of these charges will depend on the total cost of the installation and the applicable interest rate and amortization period.

The other fixed costs are due to charges for taxes and insurance. Since these are usually on the order of 1 or 2% each of the installed cost for the plant, they can be assumed at 3% for estimating purposes.

The installed cost of a particular plant, in the size range under discussion, can vary from about $5.00 per gallon per day of installed capacity for a 50,000 gpd plant, to $1.00 per gallon per day of installed capacity for a 1 mgd plant. A 50,000 gpd plant therefore costs about $250,000, installed, and a 1 mgd plant about $1,000,000, installed. These costs, for plants with performance factors from 6 to 8, include seawater intakes and steam generating equipment, but do not include costs for product water storage. Storage capacity may vary widely and is needed no matter what method is utilized to produce water at the site.

Horizontal-tube multiple-effect (HTME) plants cost less, installed, than those using the multistage flash (MSF) method, basically because the overall coefficient of heat transfer for HTME plants is more than twice as great as that for MSF plants; 1,200 to 2,000 Btu's per hour per ft.^{2} per °F. for HTME as compared to 600 for MSF. Thus, HTME plants require less tubing and result in more compact installations with smaller shell volumes. Capital costs, consequently, are less.

With heat transfer tubes in horizontal bundles, effects are low in height and arranged vertically for gravity flow of brine and condensate between effects (Figure 1). Eliminating the need for intereffect pumping this design achieves economies in capital, operating and maintenance costs, and reduced downtime for repairs and replacements.

FIGURE 1: PACKAGED 100,000 GPD 8-EFFECT HTME DESALTING PLANT

INSTALLED PLANT COST FACTORS

Factors that influence installed plant costs (capital cost) for a particular installation are listed in Table 2.

TABLE 2: INSTALLED PLANT COST FACTORS

(1) Performance factor
(2) Interest rate and amortization period
(3) Materials of construction
(4) Site costs and possible plot-size limitations
(5) Site induced costs, i.e.,
  (a) seawater intake and required structure
  (b) wharves for fuel intakes
  (c) distance and elevation from water source
  (d) distance from and mode of brine disposal
  (e) foundation or soil conditions for supporting structures
  (f) unloading facilities availability
  (g) ecological considerations
(6) Feed water composition, temperature and degree of pollution
(7) Product water conditions desired
(8) Availability of waste heat
(9) Source, capacity and location of existing electrical services
(10) Type and source of fuel and fuel storage capacity, if single purpose plant
(11) Degree of control required
(12) Turndown requirements

These considerations affect installed cost of the plant as follows:

(1) Performance factor of a distillation type desalting plant is the number of pounds of water produced per 1,000 Btu's of process steam. Defining performance factor as the number of pounds of product water produced per pound of input process steam is inaccurate, although steam has a heat of condensation of about 1,000 Btu's per pound in the range under discussion. Accurate use of the latter definition would require stating the conditions of availability of the steam.

We can build plants with a performance factor of 4 in low fuel cost areas, while in St. Croix, where fuel is moderately expensive, a performance factor of 12 or 14 might be economical. Plants having even higher performance factors can be built.

Installed cost will increase with the performance specified, which is why, in the plants mentioned earlier, costs were indicated for a performance factor of about 6 to 8. The total installation must obviously be studied to arrive at the most economical plant for the prevailing circumstances.

(2) Interest rate and amortization period influence capital costs for the plant and must be taken into consideration in determining the performance factor. If all other factors are the same, the longer the writeoff time the greater is the number of effects, i.e., higher performance factor, that can economically be built into the plant.

The interest rate and amortization period are the most important fixed-cost factors in the total cost of desalted water. For plants within the size range reviewed here, five, seven or at most ten years are usual amortization periods. Interest rates may be 8, 10 or even 12 percent.

(3) Construction materials for the plant probably offer the greatest opportunities for modifying its cost and should be taken into consideration when evaluating bids from desalting plant suppliers. Slightly higher plant costs occasioned by utilizing superior corrosion resistant materials, for example, can pay off in reduced maintenance charges and more important, especially in the size range of plants being discussed, in keeping a plant on the line.

(4) Site cost becomes significant only when land is costly. Vertically stacked HTME plants require smaller sites than that for any other distillation process and offer particular advantages if ground space is limited, as at St. Thomas.

(5) Site-induced costs include:

    (a)   Seawater intakes can take many forms, such as drilled wells, infiltration galleries terminating in a pit or sump, an open seawater intake line, or a pump mounted on a pier, wharf or float. Costs are difficult to predict and each installation must be evaluated individually.

    (b)   Wharves for fuel intakes or supplies may or may not be part of the desalting plant but, in any event, such site-support structures or means for landing fuel and supplies are needed. Hence, a portion of charges for such site-service facilities should be charged to the desalting plant. On the other hand, if water is barged in, some portion of facilities for receiving water is part of its total cost.

    (c)   Distance and elevation of the desalting plant from the feedwater source necessarily influences the cost of the desalting plant installation. Greater distances and elevations unavoidably increase piping and pumping power costs and may require larger line and pump sizes.

(d)   Distance from and mode of brine disposal influence cost although, for the size range of plants discussed here, brine disposal is seldom a serious problem. Under these conditions, brine blowdown, when mixed with condenser cooling waters, will have little biological or thermal effect on the ecology of most receiving waters.

(e)   Foundation or soil conditions for supporting structures influence plant cost. If the plant site requires the driving of piles, foundation costs will obviously be significant. In such situations, HTME plants, with small site requirements, are specially advantageous.

(f)   Unloading facilities at the site influence the installed plant cost. Packaged plants in the range of 50,000 through 500,000 gpd may require the use of specially equipped landing vessels and cranes. If the facilities for unloading a plant or transporting it to the plant site impose size restrictions, they can influence plant design as the plant may have to be built in sections and field-welded together. This is the trend, in order to take advantage of lower cost and improved quality control in the fabricating shop.

(g)   Ecological factors at the desalting plant site should be considered at all times. Laws may limit effluent temperature or concentration. Distance of discharge outfalls from shore may be a subject of regulation or code. To minimize thermal pollution, a cooling tower (or air cooled condenser) may be necessary. As always, each site must be studied and a suitable, economical design developed.

(6) Feedwater composition influences the permissible top temperature for the plant (Figure 2). Because the salts concentration at the high temperature end of an HTME plant is always less than that in an MSF plant, the HTME plant can be operated at the same temperature as the MSF plant with less danger of scaling or, conversely, at higher temperatures with equal danger of scaling.

Scaling potential is also illustrated in Figure 2. The sloping line on the left represents equilibrium conditions under which calcium sulfate anhydrite will form after a long time. Because anhydrite scale formation is strongly time-dependent, all desalting plants operate safely at temperatures and concentrations above the equilibrium line.

More important for operation is the sloping line on the right of Figure 2. Representing the temperature and concentration of seawater at which anhydrite will form instantly, this curve has been determined largely by trial and error during plant operation. At $250°F$. maximum temperature, the MSF plant, in which typical recycle concentration ranges from 52,000 to 58,000 ppm,

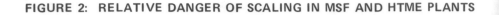

FIGURE 2: RELATIVE DANGER OF SCALING IN MSF AND HTME PLANTS

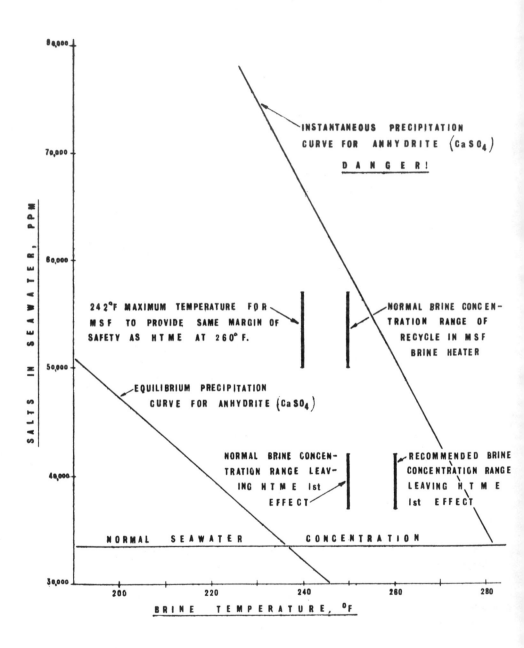

operation is very close to the danger line. Excursions in operating conditions or nonuniformity in recycle distribution in the brine heater may thus result in scale formation.

In contrast, when operating at $250°F$. brine temperature out of the first effect on normal seawater, an HTME plant will have a brine concentration ranging from 37,000 to 42,000 ppm and a margin of safety from scaling from three to five times greater than that for MSF. Operating at a $260°F$. brine temperature out of the first effect provides a wider margin of safety than for MSF plants at $250°F$. Inspection of the curves thus reveals the influence of feedwater composition at both the top and bottom of the plant.

Feedwater temperature influences plant cost from two viewpoints: firstly, the higher the seawater or sink temperature, the greater the heat transfer area required in the final condenser; secondly, higher seawater temperatures result in more costly plants since the overall available temperature difference for the plant is reduced. To illustrate the latter point, two examples follow.

If the top temperature in the plant is, in one example, $250°F$., and the condensing temperature in the final condenser is selected at $120°F$. for $85°F$. inlet seawater and a $20°F$. rise in the final condenser cooling water flow, overall plant delta T would be $130°F$. In another example, if seawater temperature is $100°F$., assuming the same approximate temperature spreads in the plant, overall plant delta T would be reduced to $115°F$.

In a ten-effect plant, the difference between an average delta T per effect of 13 degrees and 11.5 degrees is not too important. But in a 20-effect plant in a high fuel-cost area, the average delta T's per effect are $6.5°$ and $5.75°F$., respectively. With boiling point rise losses of about one degree per effect, an apparently small increase in seawater temperature can influence the required area of heat exchange surface and, consequently, plant cost. (The above delta T's per effect are stated without considering boiling point elevation losses.)

Feed seawater may be polluted by oil or chemicals, removal of which may require special techniques particularly if present in significant concentrations.

(7) Product water conditions desired influence plant cost to a small degree. The depth and type of demisters required influence cost. For potable water applications, cost additions due to demisters are small; for more exacting applications, as those requiring product water purities of 2 ppm or less, demister costs can become significant. Also, required evaporator shell volumes may increase and influence cost.

Product water discharge pressure required will influence the selection and cost of the product water pump. Product water temperatures can also increase costs by requiring a separate product water cooler.

(8) Waste heat, if available, can reduce the cost of producing water but influences

installed cost of the desalting plant only slightly. Utilization of waste heat should be considered even if its quantity or quality are limited. Supplementary firing of auxiliary and standby equipment can be engineered into a thermodynamically balanced system. Potential savings by using waste heat are obvious in the examples calculated further on in this discussion.

(9) Cost of the electric power installation for the desalting plant can be substantial, especially in a remote location. Where industry is already well developed, however, power runouts and substations near the desalting plant location may already exist. A particular advantage of the gravity-flow HTME plant is that it requires significantly less power than an MSF plant of the same capacity and somewhat less than a conventional multiple-effect plant.

(10) Type and source of fuel and required fuel storage capacity are factors that influence capital costs of a desalting plant. To produce pure water for the lowest final cost, a higher price for fuel necessitates a higher performance factor and consequently a greater capital investment.

At some locations, especially in the Caribbean, No. 6 fuel oil is unobtainable at any price because vessels with heated compartments are too large to be unloaded at the existing shallow-draft wharves. No. 2 oil must be used or costly tanker/barge landing sites installed. Even where suitable landing sites exist, a desalting plant operator may have to install a tank capable of holding three to six months supply of fuel, because the smallest quantity shipped may be 6,000 barrels, about 275,000 gallons. Fuel oil storage facility size thus becomes a function of the tanker's carrying capacity per lot delivered rather than the quantity of fuel utilized during a specific time period.

(11) Degree of control required — Costs of automating the plant represent a significant portion of total cost, especially for a small plant. Complexity of plant instrumentation and control should not exceed the capability of plant personnel to operate and maintain the system. Since most desalting plants are built in relatively less developed areas, rugged, easily maintained controls and instruments are greatly favored.

(12) Turndown — Ability to vary output simply by adjusting steam flow to the first effect is a distinctive advantage of the HTME plant. The relationship of plant cost to turndown is rather complex, since instrument ranges, liquid levels and pumps must be selected and built into the plant to accommodate expected variations in flow and capacity resulting from modifications in plant throughput.

OPERATING COSTS

Fuel costs make up by far the greatest part of the total cost of water production. The cost of fuel per 1,000 gallons of product water varies directly with the price of fuel and inversely with the plant performance factor, as explained

before. Table 3 shows approximate recent fuel prices at typical locations and
the resulting performance factors recommended.

TABLE 3: FUEL COST VS. TYPICAL PERFORMANCE FACTOR

Location	Fuel Price, ¢/gallon	Performance Factor
Bahamas	No. 2 - 22	16
	No. 6 - 17	14
Netherlands	No. 2 - 13	10
Antilles	No. 6 - 12	10
Algeria	Nat. gas - 2	3 - 4
	(Cost of gas equivalent in heating value to one gallon of fuel oil)	

Power costs are often significant at the usual sites for desalting plants. In the
United States, power may cost on the order of one or two cents per kwh, but
in the Virgin Islands, it may cost four or even six cents per kwh, especially for
on-site generation. HTME plants consume less power than either of the other
modes of distillation-type desalting plants, multiple-effect or MSF. MSF plants
require high power-consuming recycle pumps; multiple-effect plants do not.

Labor Costs — Actual cost for operating labor, including overhead and supervisory charges. Since labor costs are significant for small plants, owners should
consider the opportunities for spreading labor costs over a larger base by integrating the water plant with a power plant or other on-site mechanical equipment; i.e., in overall planning for a hotel or industrial facility.

Chemical Costs — For normal seawater applications (utilizing the HTME method),
about 2.2 pounds of 93% H_2SO_4 are needed for each 1,000 gallons of product
water. When processing normal Caribbean seawater, the cost of acid is taken
to be about 10 cents per 1,000 gallons of product water. This value is taken
since sulfuric acid may be difficult to obtain at some locations, although its
nominal cost is about 25 to 30 dollars per ton, or about 1.5 cents per pound.
Multiple-effect plants again offer economies of acid consumption since they
use less feedwater to be treated with acid than in MSF plants.

Maintenance costs, including materials and labor, are usually estimated on the
basis of an experience factor and plant size, cost and location. Maintenance
cost allowances might be 5% of plant equipment cost for a 50,000 or 100,000
gpd plant, 1.5 or 2% for a 1 mgd plant (see Table 4).

Plant availability influences product water cost and must be considered in a
rigorous estimate. Examples expressed in the equations (Table 5) demonstrate
the importance of this factor. Design considerations influencing plant availability involve selection of materials of construction, degree of sparing of pumps

and control valves, reliability of off-site utilities, and selection of superior equipment components.

TABLE 4: TYPICAL MAINTENANCE COST FACTORS

Plant Size, gpd x 10^3	Plant Equipment Cost	Maintenance Cost Factor, Percent of Plant Equipment Cost	Annual Maintenance Cost
100	$200,000	4	$8,000
1,000	$750,000	1.5	$11,250

TABLE 5: ILLUSTRATION OF INFLUENCE OF PLANT AVAILABILITY FACTOR ON COST

Case 1:

$$\frac{\text{Annual fixed costs } \$67,500/\text{year}}{100,000 \text{ gal./day} \times 365 \text{ days/year}} = \$1.85/1,000 \text{ gal.}$$

Case 2:

$$\frac{\text{Annual fixed costs } \$67,500/\text{year}}{100,000 \text{ gal./day} \times 365 \text{ days/year} \times 0.90} = \$2.06/1,000 \text{ gal.}$$

$\uparrow$

(plant availability factor)

WATER COST CALCULATION, 100,000 GPD PLANT

Water production cost factors are computed in sample calculations for a 100,000 gpd 8-effect plant located at an average Caribbean site (Example 1). Installed cost of the single-purpose plant is assumed to be $350,000, to be written off over a ten year period at 10% interest. Plant availability factor is 90%, No. 2 fuel oil costs 20 cents per U.S. gallon, and power costs 3 cents per kwh.

Desalting plant equipment cost is assumed at $200,000. Since this is a small plant in an outlying area, annual maintenance labor and material is charged at 4% of the plant equipment cost. The cost of product water from the 100,000 gpd plant studied is $5.66 per 1,000 gallons.

EXAMPLE 1: WATER PRODUCTION COST PROJECTION FOR A 100,000 GPD PLANT

Fixed Costs, 100,000 gpd

Interest and Amortization Charges:

$$\$350{,}000 \times 0.163 = \$57{,}000/\text{year}$$
$$\uparrow$$
(capital recovery factor, 10 years, 10% interest)

$$\frac{\$57{,}000/\text{year}}{100{,}000 \text{ gal./day} \times 365 \text{ days} \times 0.90} = \$1.74/1{,}000 \text{ gal.}$$
$$\uparrow$$
(plant availability factor)

Taxes and Insurance Charges:

$$\$350{,}000 \times 0.03 = \$10{,}500/\text{year}$$

$$\frac{\$10{,}500/\text{year}}{100{,}000 \text{ gal./day} \times 365 \text{ days} \times 0.90} = \$0.32/1{,}000 \text{ gal.}$$
$$\uparrow$$
(plant availability factor)

Total Fixed Costs:

$$\frac{\$1.74}{1{,}000 \text{ gal.}} + \frac{\$0.32}{1{,}000 \text{ gal.}} = \frac{\$2.06}{1{,}000 \text{ gal.}}$$

Operating Costs, 100,000 gpd

Fuel Costs:

Process Steam —

$$\frac{1{,}000 \text{ gal.} \times 8.33 \text{ lbs./gal.}}{6.5 \text{ lbs./1{,}000 Btu}} = 1.280 \times 10^6 \text{ Btu/1{,}000 gal. water}$$
$$\uparrow$$
(performance factor)

Ejector Steam —

$$\frac{215 \text{ lbs.}}{\text{hr.}} \times \frac{1{,}100 \text{ Btu}}{\text{lb.}} = 237{,}000 \text{ Btu/hr.}$$

$$\frac{237{,}000 \text{ Btu}}{\text{hr.}} \times \frac{24 \text{ hr.}}{\text{day}} \times \frac{\text{day}}{100{,}000 \text{ gal.}} = \frac{57{,}000 \text{ Btu}}{1{,}000 \text{ gal. water}}$$

Total Heat Requirement —

$$1.28 \times 10^6 \text{ Btu/hr.}$$
$$\underline{0.06 \times 10^6 \text{ Btu/hr.}}$$
$$1.34 \times 10^6 \text{ Btu/hr.}$$

$$\frac{1.34 \times 10^6 \text{ Btu/1,000 gal. water}}{(135 \times 10^3 \text{ Btu/gal. no. 2 oil}) \times 0.82} = \frac{12.1 \text{ gal. fuel oil}}{1,000 \text{ gal. water}}$$

(heating val. oil)   (boiler eff.)

Assume 12.5 gal./1,000 gal. water to allow for losses

Total Fuel Cost —

$$\frac{12.5 \text{ gal. oil} \times \$0.20/\text{gal. oil}}{1,000 \text{ gal. water}} = \frac{\$2.50}{1,000 \text{ gal. water}}$$

Power Costs:

Assume boiler	5 kw
Seawater pump at	15 kw
HTME plant	18 kw
Lighting and misc.	3 kw
Total power	41 kw

$$\frac{41 \text{ kw} \times \$0.03/\text{kwh} \times 24 \text{ hr.}}{100,000 \text{ gal./24 hr.}} = \frac{\$0.30}{1,000 \text{ gal. water}}$$

Maintenance Costs (Include Material and Labor):

$$\frac{\$200,000 \times 0.04}{365 \text{ days} \times 100,000 \text{ gal./day} \times 0.90} = \$0.25/1,000 \text{ gal.}$$

(plant availability)

Labor Costs:

Assume labor costs total $40 per day

$$\frac{\$40/\text{day} \times 365 \text{ days}}{100,000 \text{ gal.} \times 365 \text{ days} \times 0.90} = \$0.45/1,000 \text{ gal.}$$

(plant availability)

Chemical Costs:

Rounded to: $0.10/1,000 gal.

EXAMPLE 1: WATER PRODUCTION COST SUMMARY, 100,000 GPD

	$/1,000 gallons
Fixed charges	2.06
Fuel costs	2.50
Power costs	0.30
Maintenance (includes labor and materials)	0.25
Operating labor	0.45
Chemical costs	0.10
Total costs	$5.66

WATER COST CALCULATION, 500,000 GPD PLANT

A second sample calculation for a 500,000 gpd plant shows the influence of increased volume on water costs (Example 2). Site and financial conditions are assumed to be the same as for Example 1; installed cost of the 500,000 gpd plant is $700,000; desalting plant equipment cost is $500,000, and the annual maintenance labor and maintenance material charge is 2% of the plant equipment cost.

The cost of water produced in the 500,000 gpd plant utilized in Example 2 is $3.87/1,000 gallons.

Example 2 indicates that, for the above conditions, the number of effects (performance factor) should be optimized. A greater number of effects increases installed plant cost and fixed charges. But increased capital cost will be more than offset by decreased fuel consumption, and the final cost of product water will be lower.

EXAMPLE 2: WATER PRODUCTION COST PROJECTION FOR A 500,000 GPD PLANT

Fixed Costs, 500,000 gpd

Interest and Amortization Charges:

$$\$700,000 \times \underset{\uparrow}{0.163} = \$114,000/\text{year}$$

(capital recovery factor, 10 years, 10% interest)

$$\frac{\$114,000/\text{year}}{500,000 \text{ gal./day} \times 365 \text{ days} \times \underset{\uparrow}{0.90}} = \$0.70/1,000 \text{ gal.}$$

(plant availability factor)

Taxes and Insurance Charges:

$$\$700,000 \times 0.03 = \$21,000/\text{year}$$

$$\frac{\$21,000/\text{year}}{500,000 \text{ gal./day} \times 365 \text{ days} \times 0.90} = \$0.13/1,000 \text{ gal.}$$
                                                    ↑
                                       (plant availability factor)

Total Fixed Costs:

$$\frac{\$0.70}{1,000 \text{ gal.}} + \frac{0.13}{1,000 \text{ gal.}} = \frac{\$0.83}{1,000 \text{ gal.}}$$

Operating Costs, 500,000 gpd

Fuel Costs:

Process Steam —

$$\frac{1,000 \text{ gal.} \times (8.33 \text{ lbs./gal.})}{6.5 \text{ lbs./1,000 Btu}} = 1.280 \times 10^6 \text{ Btu/1,000 gal. water}$$
                    ↑
           (performance factor)

Ejector Steam —

$$\frac{750 \text{ lbs.}}{\text{hr.}} \times \frac{1,100 \text{ Btu}}{\text{lb.}} = 825,000 \text{ Btu/hr.}$$

$$\frac{825,000 \text{ Btu}}{\text{hr.}} \times \frac{24 \text{ hr.}}{\text{day}} \times \frac{\text{day}}{500,000 \text{ gal.}} = \frac{39,600 \text{ Btu}}{1,000 \text{ gal. water}}$$

Total Heat Requirement —

$$1.28 \times 10^6 \text{ Btu/hr.}$$
$$\underline{0.04 \times 10^6 \text{ Btu/hr.}}$$
$$1.32 \times 10^6 \text{ Btu/hr.}$$

$$\frac{1.32 \times 10^6 \text{ Btu/1,000 gal. water}}{(135 \times 10^3 \text{ Btu/gal. no. 2 oil}) \times 0.82} = \frac{11.9 \text{ gal. fuel oil}}{1,000 \text{ gal. water}}$$
              ↑   ↑
       (heating val. oil)   (boiler eff.)

Assume 12.5 gal./1,000 gal. water to allow for losses

Total Fuel Cost —

$$\frac{12.5 \text{ gal. oil} \times \$0.20/\text{gal. oil}}{1,000 \text{ gal. water}} = \frac{\$2.50}{1,000 \text{ gal. water}}$$

Power Costs:

Assume boiler at	20 kw
Seawater pump at	40 kw
HTME plant	44 kw
Lighting and misc.	6 kw
Total power	110 kw

$$\frac{110 \text{ kw} \times \$0.03/\text{kwh} \times 24 \text{ hr.}}{500,000 \text{ gal.}/24 \text{ hr.}} = \frac{\$0.16}{1,000 \text{ gal. water}}$$

Maintenance Costs (Include Material and Labor):

$$\frac{\$500,000 \times 0.02}{365 \text{ days} \times 500,000 \text{ gal.}/\text{day} \times \underset{\uparrow}{0.90}} = \$0.06/1,000 \text{ gal.}$$

(plant availability)

Labor Costs:

Assume labor costs total $100 per day

$$\frac{\$100/\text{day} \times 365 \text{ days}}{500,000 \text{ gal.} \times 365 \text{ days} \times \underset{\uparrow}{0.90}} = \$0.22/1,000 \text{ gal.}$$

(plant availability)

Chemical Costs:

Rounded to: $0.10/1,000 gal.

EXAMPLE 2: WATER PRODUCTION COST SUMMARY, 500,000 GPD

	$/1,000 gallons
Fixed charges	0.83
Fuel costs	2.50
Power costs	0.16
Maintenance (includes labor and materials)	0.06
Operating labor	0.22
Chemical costs	0.10
Total costs	$3.87

CONCLUSION

These examples illustrate the economics of middle-size desalting plants and suggest how the cost of product water in dollars per thousand gallons is determined. The entire analysis also explains why the question, "How much does the water cost?" cannot be accurately answered without thorough study of each factor for each installation.

Further, it should be noted that the projected costs for water can be decreased by the use of waste heat to provide steam to drive the desalting plant. Also, the foregoing projections are for high fuel cost areas and commercially realistic amortization periods; costs can be reduced by utilizing longer periods for amortization.

THE MARKET FOR DESALTING PLANTS

Raymond W. Durante
Westinghouse

ABSTRACT

Concentrated effort to improve desalting technology has been going on for over 20 years by both government and industry. The fundamental concept of converting salt water or brackish water into potable water is not only technically feasible but has a great deal of appeal as a means of effectively utilizing our natural resources.

Despite this, the widespread use of water conversion plants as a means of supplying potable water has not been accepted. In the United States, no major city relies on desalting for their supply of municipal and industrial water, and no viable commercial market exists. At the same time, projections released by the Federal government predict the production of over 2,200 mgd during the 1980 to 2000 time period.

To meet this prediction, it is obvious a considerable expansion of the desalting market must occur and the concept of utilizing desalting plants must be accepted. This paper does not examine in depth but merely suggests some of the problems which would inhibit this market expansion and presents some ideas as to how these restrictions can be overcome.

Much has been said about the potential for desalting processes as a means for implementing our supply of potable water. The subject has received considerable attention from both the industrial and government sector. Indeed, a separate United States agency was established in 1952 to provide for government sponsored research and development into means for providing potable water from the sea and other saline waters. While technology advances appear to have been made since 1952, the fact remains that today, only one United States city, Key West, Florida, depends solely on desalted seawater for the bulk of its municipal water supply. No other United States city is actively constructing a desalting plant for that purpose. True, there are listed some 750 plants worldwide which have been constructed supplying a total of 80 million gallons of water a day. But as for the United States market, other than sales to United States possessions in the Caribbean, there have been no real indications that a viable industrial market is emerging.

In a recent report prepared by A.D. Little Co. and published by the Office of Saline Water, projections of the desalting market were made to year 2020 and were broken down according to the estimated plant size. According to this report, over 500 mgd of capacity will be added in the United States for plants in the range of 1 to 50 mgd, during the time period 1980 to 2000. (This does not include an additional 1,500 mgd of plant capacity during the same time period in plants greater than 50 mgd in size listed as a separate category.)

There has been considerable controversy in the field as to the real market for very large plants of 50 mgd and greater. For the purpose of this discussion, let us look only at those plants under 50 mgd and assume that the construction of large plants is a separate problem. Such an assumption is safe to make at this point since it is probable that any large plant built in this near term time period would probably have sizeable government subsidy and, therefore, would not reflect the actual commercial market for desalting plants. It is the contention of this writer that the major commercial market will be in smaller plants.

To satisfy the projected 500 mgd, therefore, somewhere between 20 and 30 plants averaging 20 mgd would have to be built during this period. Where then in the United States is there a market for this many plants?

In new or rapidly growing areas along the seacoast where the water transport facilities are inadequate or new aquaduct and pipelines must be built, desalting is a reasonably competitive alternative to incremental water supply. This includes areas such as coastal California and the Gulf of Mexico.

The matter of water rights, however, becomes fairly critical since even though water may be available, its transfer from one place to another is usually prohibited by law, and legislation needed to change these laws is often a very lengthy and costly process. In areas where there is not potable water to start with, such as the southwest and the Caribbean, the conversion of seawater or brackish water represents an attractive alternative and consequently, a market presents itself here. Unfortunately, these areas usually do not have the economic or population base to support such projects except perhaps for smaller plants used for recreation or resort purposes.

Most of the major population centers of the East and the South have sufficient water supply which is replenished through abundant rainfall. There is a major problem emerging in these areas, however, of deteriorating quality of the water. Some municipalities have reached a point where the normal water treatment processes are no longer adequate to bring potable water up to United States health standards. Adding to the dilemma of the municipalities are the stringent environmental regulations which are imposed upon waters discharged into flowing rivers and streams.

Under the new Federal Clean Water Standards and Discharge Permit Regulations, the treatment of discharges has become a major expense factor in municipal planning. Consequently, the combination of poor quality water coming in, and the high cost of complying with pollution regulations to clean up water going out, have created a whole new set of economics for water users. These conditions appear to be driving more municipalities to the concept of closed cycle water systems and thereby, opening new doors for applications of desalting processes. This could then be the major market for desalting plants.

It should be pointed out at this time that the desalting processes usually thought of when we speak of 20 to 25 million gallon per day plants comprise the various distillation and evaporating processes and in particular, the MSF and VTE processes. True, major advances have been made in membrane technology and they could well replace the distillation processes in the future.

However, it is the general consensus at this time that for plants in excess of 10 million gallons per day, the distillation processes will still be used for some time. When considering a desalting plant to supplement the potable water supply for municipalities, one must take into account the possibility of the production of high purity water for blending purposes as well as direct consumption.

Many years ago when the Federal Water Quality Administration was formed, there was a general reluctance to consider schemes for pollution abatement which merely involved the addition of higher purity waters. In fact, they had a slogan which stated that "dilution is no solution to pollution." The basis for this thinking was that naturally occurring high purity water was too valuable to use merely as a means to dilute already polluted waters.

In the case of a desalting plant, however, we are not using naturally available high purity water but are creating high purity water from water which otherwise would have been wasted. The cost of producing desalted water can be spread out more evenly if it is blended on a 3 to 1 or 4 to 1 ratio.

In this way, one can visualize either single purpose or dual purpose (power or process steam plus desalting) plants treating waste waters and blending the high quality product with other incoming water to be recirculated through the system. Since water, unlike electricity, can be stored, it may be possible to utilize electric power generating plants during their off-peak periods to provide the necessary heat for the distillation plants and store the water for later distribution.

Assuming we have now made a case for the need for desalting plants in the medium range, (20 to 30 million gallons per day) we should now examine what must be done to make these plants commercially viable. First, the basic idea must be accepted by the municipalities and the Federal government, and then it must be demonstrated that plants such as these can be built and operated successfully. Most important, we have to examine carefully the question of the actual cost of product water.

Perhaps because the OSW charter specifically calls for large scale desalting, or perhaps because of the glamour attendant on large water projects and in particular nuclear projects, we in this country have put an inordinate amount of emphasis on large desalting projects. We have created the impression that the cost of water can be reduced through scaleup.

The early engineering studies made on projects such as the Bolsa Island project and the U.S./Mexico desalting study, projected that large scale water production could bring the cost of desalted water down to the 15 to 25 cent range and thereby be competitive with natural water supplies.

This was an honorable attempt to show the potential benefits of large scale desalting and indeed, had projects such as these proceeded, they may have achieved this goal. Unfortunately, however, they did not proceed, but the economics which were generated remained in the minds of potential water users. Consequently, the real costs actually generated from operation of their operating plants in the 1 to 5 mgd range appeared monstrously unattractive.

We in this country have been fighting a philosophy for many years which espouses that water is a free commodity or at least available at almost negligible cost. This philosophy is a result of the fact that we are blessed with abundant supplies of water and in the early days it was available for the taking. Then, as progress set in, municipal planners and managers always seemed to find ways to hide the real cost of water so that the consumer paid not the actual cost but a subsidized cost.

Perhaps then, one of the first things we must do to establish a strong market for desalting plants is to embark on an education program which brings to light the actual cost of water as related to its worth as a commodity and not some predetermined level which we feel it should be.

The cost of manufactured water from a desalting plant is like other manufactured items, sensitive to many things. One of the most misunderstood values in calculating the cost of water is plant capacity. Usually an assumption is made that the plant will operate for at least 85% of the time on a continuous basis and water costs worked out from that point.

This is a throwback to the electric utility industry which has thousands of man-years operating experience to justify high plant capacities. There have been very few desalting plants that have operated for any length of time anywhere near their rated capacities and therefore, a more realistic approach to this figure must be taken. Also, the operations of a desalting plant are far more critical than we like to admit. The basic concept of boiling saltwater and recovering fresh water is as old as man himself. Therefore, we tend to minimize the need for sophisticated control and operating procedures and, more importantly, trained operators. An unfortunate paradox exists that most of the areas in the world utilizing desalting plants are also in areas of relatively low labor skills.

It is difficult to get and keep highly trained operators for desalting plants. The same fact holds true for field erection and construction labor. No area of costs has skyrocketed as rapidly as field labor, especially when one requires the high quality type needed for the assembly of large evaporator units used in desalting plants.

Many companies find it necessary to transport entire work crews and supervisors as well as engineering management personnel to the site to remain there for the duration of the construction project in order to insure the proper erection of a desalting plant. With U.S. labor rates, such a procedure is almost prohibitive and may be the dominant factor in driving U.S. desalting firms out of the international market.

Here in the U.S. a new concern generated by the influx of environmental regulations and the energy crisis is the high cost of fuel. Fuel costs today are far from stable and can change significantly almost overnight. The cost of fuel, is of course, highly critical to the water costs, and as we become more dependent upon imported low pollutant fuels or exotic pollution abatement schemes, the cost of desalted water may increase greatly. The same is true for chemical costs and other materials which are supplied on a continuing basis.

What can we do then to help create a viable commercial industry to supply the 500 million gallons per day desalting plants or, if you will, 25 to 20 million gallons per day plants throughout the U.S. by the year 2000? It has been said that research is the basis for all productive enterprises and certainly we should continue our basic and applied research in all areas of water conversion.

It may be, however, that there will be no major breakthrough which would drastically reduce the cost of water and the installed cost of a desalting plant. Instead we might be able to reduce these costs by examining some of the present plant concepts and attacking directly the problems which now exist and which are preventing desalting from emerging as a viable business.

In the past 5 years, there have been a number of new processes which promise to provide the solution to viable desalting. The MSF/VTE, vapor compression, freezing process, vertical tube, horizontal tube, all these various modes take turns at being the answer to our problems. Considerable funds by industry and government are spent in developing and promoting these concepts.

If there is to be a subsidy to the desalting industry, perhaps it might best be directed towards the construction of plants using the technology that has been in existence for a number of years. It may be wiser to demonstrate the technical feasibility of a given process and then follow with its continued economic

improvement rather than try and prove economic acceptability before attempting to build a plant. It may be more prudent to standardize the engineering
concepts and build a number of plants which can operate under real life conditions and demonstrate conclusively whether or not potable water can be made
from brackish or seawater and used beneficially.

If we had available today desalting plants which were simply constructed, perhaps even of modular construction, which were simply operated, tape controlled
or programmed, and which were reliable, to run at a minimum 60% load factor,
they would receive careful consideration from water planners regardless of the
cost of the product water.

Commodities such as water, land and air have one important thing in common.
There are no substitutes for them. Water, in particular, can no longer be used
once and thrown away. It must be reused and recycled. Bodies of water which
contain pollutants or salt cannot be written off as useless but must be considered as a potential resource in all of our planning.

There is a good chance that if we could demonstrate on a fairly wide scale the
technical and programmatic feasibility of water conversion, the economics
would fall into line as improvements are made. If we procrastinate and continue to look for the ideal solution and the ideal economics, we may never get
a plant built.

CHEMICAL PROCESSES BASED OPERATIONS THEORY AND PRACTICE

Irwin R. Higgins
Chemical Separations Corporation
Oak Ridge, Tennessee

ION EXCHANGE

Ion exchange is a reversible chemical reaction between a solid and liquid phase containing bound groups that carry an ionic charge either positive or negative, in conjunction with free ions of opposite charge that can be displaced. The exchange of ions is a phenomenon, universally spread throughout nature.

About 1850, Harry S. Thompson and John T. Way, two English agronomists, noticed a stoichiometric exchange between ammonium and calcium in soils. In 1876, the exchangers were identified as crystalline alumino-silicates and the displacement of the cations was shown to be stoichiometrically reversible. Various natural minerals, called zeolites, were found to have exchange characteristics suitable for the softening of water.

In 1904, the first artificial alumino-silicates were synthesized by a German chemist. These artificial zeolites had effective capacities two or three times those of the naturally occurring minerals. Most ion exchangers currently in large scale use are based on synthetic resins, usually polystyrene, copolymerized with divinyl benzene (to provide the requisite amount of cross-linking) in bead form. They are permeable only at molecular dimensions and their superficial physical structure is not affected during the ionic "swap-out."

The conventional procedure in ion exchange is to pass the solution to be treated through a bed of resin beads which has a limited exchange capacity. The process essentially is a batch type operation in which the resin, upon depletion, is regenerated. The most frequently used method of fluid-solid contact for sorption operations is in columnar units, or "fixed beds", with the resin closely packed. A careful examination of the performance of a fixed bed ion exchange column with cocurrent flow would reveal that a large fraction of the ion

exchange resin is not serving a useful purpose during the major portion of the
operational or service cycle. At the mid-point of the service cycle in a fixed
bed, the inlet porticn of the column is completely exhausted while the predominant part of the outlet portion of the column still has available exchange sites.

This indicates that only a segment of the resin bed is active in the ion exchange
process. Therefore, those segments of the fixed bed that are not utilized could
be continuously removed leaving only the portion of the bed that is actively
participating in the ion exchange reaction.

The intrinsic value of ion exchange encouraged many attempts to improve efficiency and capacity of the process by developing a continuous system. The
first practical continuous ion exchange system was produced by Chem-Seps
Technical Director, Irwin R. Higgins, in the early 1950's during his work at the
Oak Ridge National Laboratory.

The Higgins system is based on countercurrent flows of fluids and resin within
a continuous loop (see Figure 1). The various sections of the loop are separated
by valves to isolate the sections performing different functions simultaneously.
The resin is moved in small increments by the hydraulic ram action.

All solution flows are countercurrent to the movement of resin. The contaminants are removed in the process cycle and the resin is moved during the pulse
cycle. During the process cycle the influent enters through a set of distributors,
moves downward and leaves through a set of collectors. Simultaneously, the
regenerant enters through valve **K** and leaves through valve **E**. The rinse enters
through valve **I** and its duration is controlled by an interface controller. In
the process cycle valves **B**, **C**, and **D** are closed, isolating all the sections.

The duration of the process cycle ranges between 4 to 10 minutes, depending
on the contaminants to be removed and is controlled by a timer, if the contaminants do not significantly vary, or an analyzer controller in the effluent if
the contaminants change over relatively short duration.

Resin is dropped into the pulse vessel through valve **A** simultaneously with the
removal of contaminants in the treatment section, regeneration of the loaded
resin in the strip section, and rinsing of regenerated resin in the rinse section.

At the end of the process cycle, all flows going into and out of the unit are
shut off. The valve **A** closes and valves **B**, **C**, and **D** open in proper sequence.
A hydraulic pulse of water enters through valve **H** transferring resin around the
loop a preselected distance. This operation moves exhausted resin out of the
treatment vessel replacing it with regenerated resin while pushing loaded resin
from the pulse vessel into position for regeneration.

The total elapsed time required for sequencing valves and the movement of resin is
20 to 30 seconds. At the end of the pulse cycle, valves **B**, **C**, and **D** close and valve
A opens. The solution valves open initiating normal steady state operation.

FIGURE 1: SCHEMATIC DIAGRAM OF CHEM-SEPS CONTINUOUS COUNTERCURRENT ION EXCHANGE CONTACTOR

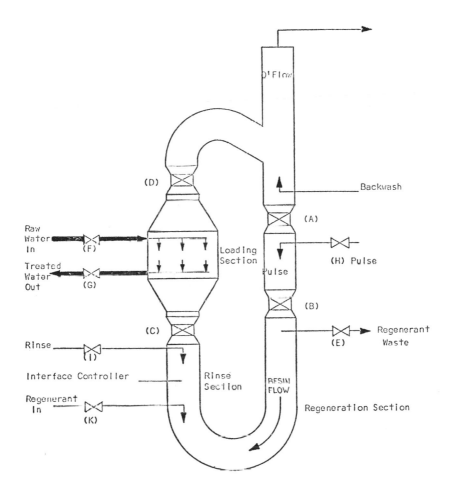

Valve Positions During Cycles

Run Cycle		Pulse Cycle	
Valves Open	Valves Closed	Valves Open	Valves Closed
A,E,F,G,I,K	B,C,D,H	B,C,D,H	E,F,G,K,I,A

The high operating efficiency of Chem-Seps continuous systems offers significant advantages. Dramatic savings on resin regenerant result from countercurrent flow of rinse and regenerant with a uniformly high quality of treated effluent. The downflow of feed overcomes the minimum flux requirements in upflow systems.

In the downflow system resin drops in the treatment vessel by gravity. The maximum recommended flux is 60 gal./min./sq. ft. This limitation has been imposed strictly because of the pressure drop requirements. However, there are no limitations on the minimum flux. Another advantage of downflow is that the liquids with higher specific gravity are always located at the bottom part of the loop, thus facilitating precise interface control with concentrated wastes.

The objective of this discussion, henceforth, will be to outline successful experiences with continuous countercurrent ion exchange focusing on some unusual or nonstandard ion exchange techniques rather than the standard "buy" and "throw away" of regenerant wastes.

To accomplish this objective, examples of large scale continuous ion exchange plants will be cited to prove the validity of the process. At the same time, specific features of ion exchange technology will be viewed from Chem-Seps experience. The following categorical breakdown will be followed to insure a proper regard for the flexibility and versatility of the system:

(1)   Industrial and Municipal Applications of
Continuous Countercurrent Ion Exchange
to Demonstrate the Practicality and Economics.

(2)   Prevention of Scale in Seawater Distillation
Units by the Removal of Calcium.

(3)   Partial Demineralization of Acid Mine Drainage
Water at Low Regenerant Cost.

(4)   Popper Ion Exchange Process for Desalination.

(5)   Industrial Waste Treatment: Chromate Removal
and Recovery System for Pollution Abatement.

SUMMARY

There are over 70 continuous countercurrent ion exchange units designed by Chem-Seps installed around the world demineralizing, dealkalizing, and softening water. Chemical efficiencies and therefore lower operating costs have been the principal reasons for installing continuous countercurrent ion exchange systems.

A prime example of successful large scale continuous ion exchange equipment is the water softening installation at Alameda County, California. The use of Hub type distributors and Johnson well screens has made it possible to handle the high flow requirements in this municipal application. Significant also to this operation is the high regenerant efficiency on a continuous 24 hour day-by-day basis.

An interesting aspect of the water demineralization system installed in Brandon, Manitoba, is the unique regeneration system and blending of wastes to form a fertilizer product. This feature eliminates the highest operating cost while giving a useful product.

The most prominent ion exchange application in the saline water program, as demonstrated to the Office of Saline Water, is the pretreatment of seawater to prevent scale formation by the removal of calcium in distillation equipment. The attractiveness of this system is the use of evaporator blowdown, normally considered a useless waste, as a softening regenerant at virtually no cost.

The treatment of Smith Township mine drainage water demonstrates the capability for substantial reduction in regenerant consumption by the use of sulfuric acid and lime slurry on a continuous ion exchange system. The reduction in sulfuric acid is accomplished by mixing strong and weak base resins to fully utilize both hydrogen ions in the sulfuric acid. The Sul-biSul anion treatment is an avenue to substantial reduction in operating costs by the successful use of lime slurry as a regenerant.

The Popper process is attractive for brackish waters because lime is the only regenerant filling the role of both sulfuric acid and lime in the more conventional demineralization applications.

There is now a specialized cooling tower recycle system that allows the use of preferred chromates as corrosion inhibitors without the problem of chromate destruction and disposal. The low operating cost of this system is achieved by the advantages of continuous and countercurrent ion exchange technology.

One can readily see the variations possible in devising flowsheets to fit the increasing demand in both the water treatment and chemical process industries. Some further examples of chemical process applications are: Magnesium removal from phosphoric acid to prevent afterprecipitation in the production of "super-acid"; recovery and treatment of certain plating wastes for cost reduction and pollution control; extraction of uranium from ore pulp and demineralization of thin sugar juice to increase yield and reduce scaling of heat exchanger.

The wide variety and cost saving features of these applications are made possible by the characteristics of continuous countercurrent ion exchange as employed and pioneered by Chemical Separations Corporation.

Industrial and Municipal Applications of Continuous Countercurrent Ion Exchange

There are vast differences between the operation of small scale ion exchange equipment compared to large scale high throughput units. The success or failure of large scale applications of ion exchange depend on the engineering and mechanical features associated with the equipment. Chemical Separations has overcome the problems pertaining to large scale equipment in a 25 million gallon per day water softening installation in Alameda County, California (see Figure 2).

FIGURE 2: ALAMEDA COUNTY WATER DISTRICT WATER SOFTENING PLANT, FREMONT, CALIFORNIA

The installation consists of four units, each with a 96" diameter treatment vessel and a 30" diameter loop. To achieve the proper distrubution of high flow influent water, Hub Type distributors and collectors are utilized. Johnson well screens are used to contain the resin in the units. The Alameda County units are now softening 250 ppm hardness well water to less than 5 at 11,200 gpm.

The treated water is blended with 9 mgd of raw water to yield final blended effluent containing less than 85 ppm total hardness. Significant in this municipal operation is the high regenerant efficiency of less than 20% excess over stoichiometric.

It is possible on a 4" diameter pilot unit, to take 300 ppm hardness water down to zero with the stoichiometric consumption of salt. To duplicate this type of performance in large scale equipment is an achievement of considerable significance.

Many industries in their quest for better products and high output capacities have found the increasing necessity for high purity water. Since distillation cannot achieve the quantity or quality of water required, manufacturers and public utilities have turned to ion exchange for low cost-high purity water.

Usually the largest single cost in operating an ion exchange demineralizer is the cost of chemicals used for regeneration. A conventional fixed-bed ion exchange demineralizer will be found to use a 100% or more excess of regenerant chemicals over the stoichiometric amounts.

The countercurrent flow efficiency of continuous ion exchange has given a reduction of regenerant consumption to within 10 to 25% excess over stoichiometric amounts.

An example of a continuous demineralization plant installed in a fertilizer complex in Canada includes a strongly acidic cation unit followed by an atmospheric decarbonator with a weakly basic anion unit and a strongly basic silica polisher (see Figure 3).

The untreated feed water to this system contains from 1,200 ppm total dissolved solids (TDS) to as high as 1,800 ppm TDS. Approximately 65% of the TDS is sodium with an effluent requirement of 5 ppm TDS.

The real aim of the system is to maintain the chloride content in the effluent to 3 ppm (sodium removal is over 99.4% and is reduced from 500 ppm to less than 3 ppm.

The cation unit is regenerated with 10 to 12% nitric acid. This acid was selected because it is an "in-plant" product with a relatively low cost. It is of interest to note that the excess nitric acid may be piped to the blending plant so that the nitrate valves are actually calculated as part of the product of this fertilizer complex.

The regenerant for the weakly basic anion unit is 3 to 5% ammonia as ammonium hydroxide solution. Ammonia is also a plant product, and thus available at low cost. If full credit is taken for regenerant chemicals, all other costs for this 1,500 ppm water are less than 5 cents per 1,000 gallons.

FIGURE 3: TYPICAL LAYOUT FOR DEMINERALIZATION

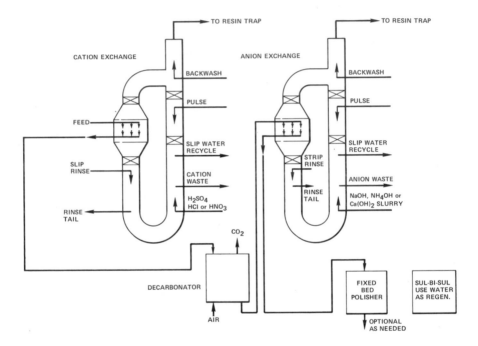

Prevention of Scale in Seawater Distillation Units by the Removal of Calcium

Removal of calcium from seawater allows an evaporator to operate at high temperatures, thus greater efficiency, without calcium sulfate scaling. Evaporator blowdown, normally considered a useless waste, serves as a valuable softening regenerant at no cost. In the softening ion exchange reaction, the effectiveness of the sodium ion for removing calcium-magnesium, increases as the square of the sodium ion concentration. Therefore, any concentration of seawater is beneficial, with the degree of calcium removal increasing with increased blow-down concentration.

In 1961, Chemical Separations Corporation conducted an application study pro-gram to determine if efficiencies could be gained by the use of continuous countercurrent ion exchange for removal of calcium from seawater (to prevent scaling of distillation equipment) compared with the conventional fixed bed equipment pioneered by Dow Chemical Company.

Using synthetic seawater and a simulated evaporator blowdown of 5, calcium removal of over 95% was achieved on a one-inch laboratory unit. As a result of laboratory studies, a demonstration unit was constructed for the Office of Saline Water (OSW) installation at Freeport, Texas (see Figure 4). This is an "upflow" contactor and not the present improved "downflow" configuration described earlier.

The evaporator at Freeport is a long tube vertical (Stearns-Rogers) with a capacity of one million gallons per day; whereas, the capacity of the continuous ion ex-change contactor is 60,000 gallons per day. These studies demonstrated over 65% removal of calcium with less than 40% removal of magnesium.

Channelling, inefficient regeneration, wasted resin inventory for off-stream re-generation, low throughput per square foot of resin area, and poor distribution of feed, brine and rinse solutions across the resin bed were not present in the Chem-Seps moving bed, countercurrent system as noted in fixed bed units.

On recommendation by Chem-Seps, further tests were made at OSW's demon-stration facilities at Wrightsville Beach, North Carolina, with an evaporator hav-ing the same capacity as the calcium removal unit.

Seawater was fed to the unit at a rate of 25 gpm while softened water was fed to a Baldwin-Lima-Hamilton evaporator to give a blowdown factor of 2. This regenerant was sufficient to render a 70% extraction of calcium. All process water was softened recycled seawater with no other water consumption. With 70% removal of calcium, the evaporator was operated for 200 hours at 275°F. without scaling. With simulated blowdown factors of 3 and 4, calcium removal was 85 and 90%, respectively.

Normally, once calcium sulfate scaling begins, it accelerates necessitating evapo-rator shutdown and a difficult clean-out process. However, Chem-Seps demon-strated that as scale begins to form it can be dissolved by simulating a higher blowdown factor to increase the degree of calcium removal. This particular ob-servation, though demonstrated, needs further study and full investigation. In the saline water program, the preremoval of hardness is applicable to reverse os-mosis and electrodialysis since these systems produce a blowdown similar to an evaporator thus sufficient for ion exchange regeneration. The effectiveness of this ion exchange application lies in the fact that there is no costly purchase of regenerant chemicals.

FIGURE 4: SCHEMATIC DIAGRAM OF CHEM-SEPS CONTINUOUS
CALCIUM REMOVAL

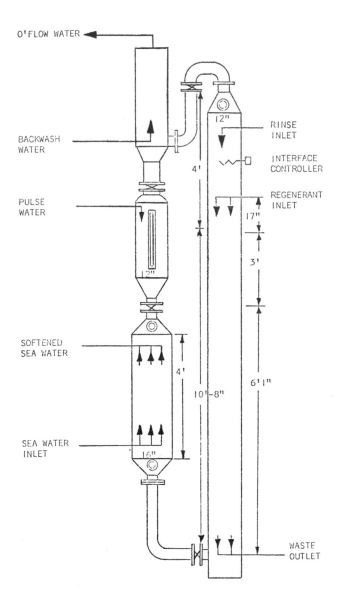

Partial Demineralization of Acid Mine Drainage Water at Low Regenerant Cost

There is now available a low cost ion exchange system for demineralization of brackish water. The method reduces regenerant consumption by utilizing strong and weak acid resin in the cation step and Sul-biSul process on the anion step. Chem-Seps has installed a demineralizer for treatment of Smith Township, Pennsylvania, mine drainage water featuring these innovations with sulfuric acid and lime as regenerants.

Past experience in the ion exchange industry has been to use 100% or more excess sulfuric acid when using strong acid resin for demineralization. Chem-Seps has demonstrated on 1,000 gallons of actual Smith Township water a substantial reduction in regenerant consumption by mixing equal parts of strong and weak acid resin to fully utilize both hydrogen ions in the sulfuric acid. The method is based on the greater efficiency of the sulfuric acid in regenerating the weak acid resin.

Consequently, the intent is to promote the greatest possible sorption of calcium and magnesium by the latter. This is achieved through a greater natural affinity resulting from the transfer of hardness ions from the strong to the weak acid resin in continuous countercurrent ion exchange. Sulfuric acid usage in the cation step of the run was less than 5% over stoichiometric, with hardness reduced from 1,100 to 150 ppm.

To complete the demineralization operation, Dow's Sul-biSul or "Site-Sharing" process is used in the anion step to give a final product water of pH 6. Sul-biSul demineralization is dependent on the two-stage ionization of sulfuric acid to sulfate ion ($SO_4^=$).

When a sulfate form anion exchange resin which has been rinsed with neutral or alkaline water is contacted by an acid solution, conversion of the sulfate ion to bisulfate occurs and half of the resin's available capacity is freed to react with the acid solution. The sulfate ion occupies two exchange sites of the anion exchanger, bisulfate only one.

When the resin bed is exhausted, it is "regenerated" by a wash of water, converting bisulfate ions back to sulfate ions and discharging sorbed acids. A stoichiometric amount of lime slurry is used in the regeneration step to neutralize the free acid and to reduce the waste volume by accelerating conversion. The use of low cost lime slurry for regeneration is more than just a substantial economic saving. It demonstrates that with Chem-Seps single loop design anion resin in the sulfate form may be effectively regenerated with lime slurry without precipitation of calcium sulfate within the moving bed.

Popper Ion Exchange Process for Desalination

While the Sul-biSul demineralization is feasible only in high sulfate waters, the Popper process is applied to brackish water high in chloride and low in sulfate.

The desalination process consists of two CCIX units: The first a mixed bed with cation and ion resins, the second with only cation resin (see Figure 5).

FIGURE 5: BRACKISH WATER VERSION OF POPPER PROCESS AS ADAPTED TO CHEM-SEPS CCIX

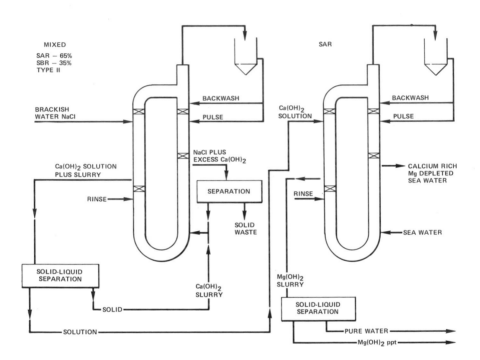

The brackish water high in sodium chloride is fed to the mixed bed unit with the resin in the calcium and hydroxide forms. Sodium and chloride are sorbed on the resin producing a calcium hydroxide solution or slurry. The calcium hydroxide is in solution if the mineralization is less than 2,000 ppm and in slurry with a mineralization greater than 2,000 ppm. A lime slurry is used to regenerate both cation and anion resins in the mixed bed. Chemically, the reaction is:

$$\begin{matrix} CaR \\ \\ ROH \end{matrix} + NaCl \rightleftharpoons \begin{matrix} NaR \\ \\ RCl \end{matrix} + Ca(OH)_2$$

The calcium hydroxide effluent from the mixed bed unit is then fed to the cation unit. This resin is in the magnesium form regenerated by seawater. As the calcium hydroxide is fed through this unit, it is converted to the more insoluble magnesium hydroxide which settles out to give a high purity water.

$$Ca(OH)_2 + MgR \longrightarrow CaR + Mg(OH)_2$$

This process enables an 80% sodium chloride removal with a near stoichiometric consumption of lime. The attractive feature of this application is the necessity of only one regenerant, lime, at low cost.

Economically this is more attractive than the low cost regenerant consumption of sulfuric acid and lime in the more conventional demineralizer. The process is made feasible by the ability to handle a lime slurry with no plugging.

At the present time, Chem-Seps and Aqua-Ion companies are negotiating a contract study with the Office of Saline Water to further develop the process. A 3 mgd installation is proposed to treat 1,000 ppm sodium chloride water. The estimated cost with lime at $20/ton and no recovery, capital or operating cost comes to about 10 cents/1,000 gallons.

Industrial Waste Treatment: Chromate Removal and Recovery System for Pollution Abatement

Chem-Seps now has a specialized cooling tower recycle system that allows the use of preferred chromates as corrosion inhibitors without the problem of chromate destruction and recovery. The advantages of continuous countercurrent ion exchange have been applied to the Nalco Chemical Company system for chromate recovery.

The low operating cost of this recovery system is the attractive feature. Operating costs based on chemicals and resin used in the Chem-Seps system compared with conventional fixed bed ion exchange units indicate the superiority of continuous countercurrent ion exchange. On a per pound cost of recycled chromates, Chem-Seps' cost per pound is 8.5 cents while fixed bed cost is 18.5 cents per pound.

The system itself requires a single continuous ion exchange unit with weak base anion resin (see Figure 6). Sodium hydroxide is the primary regenerant with near stoichiometric consumption based on the recovered chromates.

The system allows cooling tower operation at low cost with emphasis placed on pollution control.

FIGURE 6: COOLING TOWER CHROMATES RECOVERY

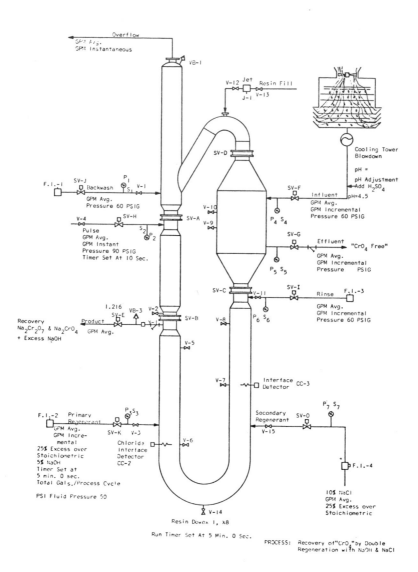

Resin Dowex 1, X8

Run Timer Set At 5 Min. 0 Sec.

PROCESS: Recovery of "CrO$_4$" by Double
Regeneration with NaOH & NaCl

DESALTING PLANT OPERATOR TRAINING

Bruce M. Watson
W.R. Grace & Co.

ABSTRACT

Training requirements for desalting plant operations personnel will vary widely with: (1) facility (power station, industrial complex, isolated resort hotel), (2) labor pool (educational level, skills and relevant experience of available manpower), (3) instruction (by contractor personnel, existing facility staff or government trade-training programs) and (4) level of responsibility for which training is intended (engineer-in-charge, shift foreman or operator).

These elements, together with other important human influences of language, regional politics, weather and customs are discussed in this lecture on personnel selection, familiarization and training for safe and efficient desalting plant operation.

INTRODUCTION

Just as for an electrical utility, the most desirable image for a desalting plant is low profile, i.e., unobtrusiveness and faithfulness. Desalting plant shutdowns, caused by leaking tubes or malfunctioning pumps, may not result in such spectacular and immediate consequences to the consumer as a power blackout, but when water storage levels drop to the service pump suctions, repercussions can be just as painful.

Equipment deficiencies must bear the blame for many of these situations. However, well-trained and supervised operators can limit these crises to manageable proportions by early recognition of trouble symptoms and appropriate action.

The purpose of this lecture is to outline the training necessary to achieve this end, and to meet the safety standards expected of any plant comprising steam, chemicals, electric power and rotating machinery.

Factors Affecting Scope of Recruitment and Training

Training must obviously differentiate between, for example, an experienced steam plant operator vs. a youthful secondary school graduate; between personnel destined to man a shift under the closely supervised environment of an oil refinery, and those expected to be responsible during a night shift at a remote island hotel.

Plant size, per se, is not an important factor. The most significant criterion is the degree to which a water supply is dependent on the operator's knowledge, experience and integrity. ·

Facility Factors

Even if it is a first, the desalting plant added to refinery or utility has the inherent benefit of an existing trained supervisory, operating and maintenance labor force. Basic background skills, i.e., steam and electrical system operation, pipe-fitting, instrument maintenance, welding, machining, etc., being immediately available, a minimum of extra training effort is required. Additional operating manpower, if indeed any is necessary, is readily drawn from the labor pool already provided for.

At the other extreme, operating staff required for a remote, single-purpose facility (water only, or water and power) may need to be trained from scratch, particularly if there is no local industry to draw from. Such training will fall into two categories:

(1)   technical fundamentals (reading instruments; recording data; understanding of units, component functions, system operation), and

(2)   equipment familiarization (design capabilities, starting and stopping, trouble-shooting, safety features and routine minor maintenance procedures specific to the plant design).

Maintenance skills, such as those referred to above are another matter, and must be sought from outside if necessary.

Labor Pool Factors

Again, very wide regional differences exist between labor forces available for recruitment, and thus between training methods applicable. One region, for example, Key West, has a substantial pool of highly skilled retired naval personnel to draw from.

Another, let us say in the Arabian Gulf, must rely on a local labor force which, however well motivated, is basically untrained even in the technical fundamentals normally considered a prerequisite for employment elsewhere.

Language differences can appear to pose great difficulties to construction, start-up and operation of a distillation facility. Local language may differ from the "technical" language of the region, which may differ again from the native language of the general contractor and again in earlier times may have differed from that used in literature, drawings and instruction books provided by each component plant manufacturer. The following table illustrates the variety of national tongues involved in recent installations:

Plant Location	Jidda S. Arabia	Las Palmas Canary Is.	Arzew Algeria	Curacao	Doha Qatar
Native Language	Arabic	Spanish	Arabic	Papiamente	Arabic
Technical Language	English	Spanish	French	Dutch	English
General Contractor	Dutch	Dutch	French	Dutch	French
Boiler Mfr.	Japanese	Spanish	–	–	–
Turbine Mfr.	German	German	–	–	–
Distillation Plant Mfr.	U.S.	Dutch	English	English	French

Actually, few problems arising during construction and startup are directly attributable to language differences per se since the technical language is recognized as common. National and political attitudes do clash at times but the greatest source of friction in technologically developing countries will be real or implied impatience and criticism by contractor personnel, vs. resentment and bruised feelings by local nationals involved.

Instruction Available

It might seem logical, and is often so specified, that the plant manufacturer's field erection and startup staff undertake training of local personnel for operation. Yet the best field installation engineers are sometimes the most impatient and inarticulate of teachers.

Nevertheless, few equipment purchasers are willing to bear the extra cost of importing an instructor especially qualified for the job and training of sorts does proceed.

Some governments, notably Kuwait, provide special courses for utility operators and maintenance personnel as well as engineers. This approach has special merit in that it can be a natural adjunct to trade-training programs, where there are usually public funds available and where direction and coordination of public works construction and trianing in developing countries is much more centralized. However, no program, or formalized course of instruction, can substitute for the on-the-job training by an experienced staff. Except where political hiring pressures may apply, (and they do), candidates are required to have some basic technical fundamentals or experience.

Level of Responsibility

The prospective engineer-in-charge will obviously need only familiarization with the specific process details and equipment supplied, both for maintenance and operation supervision.

For larger facilities, where a shift foreman is required, prior experience is also assumed and only system startup, shutdown and control procedures need be assimilated which are specific to the installation. Usually, recruitment standards for both these functions limit candidates to those having enough initiative and maturity (as well as technical experience) not to require formal instruction.

The supervised shift operator candidate may be relatively unskilled, not particularly a self-starter but willing to learn and take direction.

The unsupervised shift operator candidate, who must start with at least some familiarity with steam and electrical machinery, will normally be responsible for a complete plant, (although a separate boiler operator would also man the same shift). For this reason, his training may need to be somewhat more rigorous than for any of the above categories.

MANPOWER SOURCES, REQUIRED BACKGROUND AND JOB SPECIFICATIONS

Manpower Sources

In order of preference, ideal sources of experienced operating staff are as follows:

> Other distillation plant facilities
> Marine engineering (steam)
> Marine engineering (IC engine)
> Refinery operations
> Other chemical process industry operations
> Stationary engineering
> Municipal water treatment

Obviously, to obtain men already employed at these sources, some thievery is implied, with due consideration given to the importance of a good neighbor policy where the region is small. It is equally evident that under such circumstances, former employees may be more readily checked for suitability, except perhaps ex-marine types. (The author is one, and thus feels free to comment!) These men are usually well qualified technically, but let the remote resort hotel operator beware of the few misfits who do decide that tropic nights and cheap rum are just what is needed after all that sea time. In summary, the categories of experience considered pertinent are as follows:

steam generation and distribution (boilers; heat exchangers; valves; traps; pressure, temperature and level regulators; instruments and safety features)

pumps, compressors and vacuum equipment

electrical machinery, transformers, switch-gear and controls; instruments and safety features

internal combustion engines, air conditioning and refrigeration

The Jobs and the Men to Fill Them

Let us now examine the requirements for desalting plant operations and the appropriate background experience and training needed by the staff. Since job titles are ambiguous, and vary from facility to facility, let us clarify. For the purposes of this lecture:

"Engineer-in-charge" — refers to the man having top technical responsibility for the desalting plant's efficient operation; he may be only a lowly process engineer in a refinery, a hotel chief engineer or a power plant assistant superintendent, or utilities foreman.

"Shift Supervisor" — refers to the shift man having the greatest technical responsibility for the plant, i.e., the individual who would direct (or execute) starting, stopping and adjustments. Where there is only one man per shift, he could be required to carry the same responsibilities as listed for the Shift Supervisor. On the other hand, his duties may not extend beyond emergency shutdowns and calls for help.

"Shift Operator" — refers to one or more shift men working under shift supervision.

Job Descriptions

Engineer-in-charge: For the specific plant he must gain complete knowledge of:

(a) plant operating principles and the design and cost criteria which set the pattern for its capacity, efficiency and number of stages, temperature range and materials of constructon;

(b) unit and system function and corresponding design features; heater; stage condensers; flash chambers with vapor scrubbers and interstage sealing arrangements; seawater, brine recycle and distillate pumps; vacuum system with after-condensers and drains;

seawater intake screening and trash removal; makeup filtration, chemical addition, decarbonation and deaeration;

(c)   local and remote controls of steam pressure and desuperheating, brine temperature, stage distillate and brine levels;

(d)   local and remote instrumentation for pressure, temperature, flow, level, conductivity and pH;

(e)   principles and practice of chemical treatment; dosage calculations for solution makeup and injection rate control;

(f)   principles and practice of corrosion control; protective coatings; cathodic protection; inspections; layup procedures;

(g)   startup, shutdown and operational "tuning" for best efficiency; emergency shutdown procedures;

(h)   maintenance of all mechanical and electrical systems.

Shift Supervisor: Except as his individual initiative takes him, the shift supervisor does not require nearly the depth of design and construction detail as does the engineer-in-charge. On the other hand, he must be thoroughly competent to supervise the following activities:

(a)   plant startup procedures and adjustment to specified performance standards;

(b)   stabilization of upsets, e.g., after temporary loss of steam or electrical power;

(c)   trouble shooting, e.g., for loss of vacuum, interstage imbalance, salting over, abnormal heater fouling;

(d)   plant shutdown procedures, normal and emergency;

(e)   data acquisition and its interpretation, i.e., ensuring clarity and accuracy of log entries, including recognition of careless or bogus readings vs. instrument deviation or malfunction;

(f)   use and understanding of pH, conductivity and density measurements, and makeup and dosage of feed treatment chemical solutions and intake chlorination;

(g)   enforcement of safety codes and regulations;

(h)   instruction of operator trainees.

Shift Operators (supervised): Under direction, the shift operator (or operators) must perform a variety of supernumerary, but necessary duties as:

(a)   recording of instrument readings, local and remote reading thermometers, pressure and vacuum instruments, indicating and integrating flow meters, liquid levels, pH meters and salinity indicators; he must therefore understand units, be able to interpolate, and to distinguish between readings from inoperative instruments and actual system deviations;

(b) periodic rounds of remote equipment, i.e., intake travelling screens, seawater supply pumps, chlorination equipment, product water storage tanks, service and transfer pumps; inspection of rotating equipment (bearing and motor winding temperatures, stuffing box leakage) and lubricating routines.

(c) look out for developing problems and hazards, such as leaks, vibration, etc.

TRAINING SYLLABUS

Subject Matter

For convenience, lecture topics on the distillation facility may be subdivided as follows:

Lecture 1: Principles of evaporation and heat recovery at decreasing pressure and temperature:

> single effort and multieffect boiling or
> single stage and multistage flash
> > once through
> > brine recirculation

Simple line diagrams showing approximate temperatures and flow rates are extremely helpful for these lectures. Even though design calculations are not made or taught, the numbers will give perspective. The lecture should conclude with study of the actual plant flow diagram.

Lecture 2: Stage (or effect) mechanical construction arrangements, featuring:

> boiling or flashing region
> vapor/liquid separation
> heat transfer tubing and waterboxes
> interstage or intereffect flow control orifices
> gas venting
> materials of construction

For instruction, plant drawings (preferably extra prints) may be marked up with colored chalk or pencil to highlight and simplify.

Lecture 3: Pumps and vacuum equipment. Each item should be located by reference to the diagrams used in Lecture 1.

> seawater supply pumps
> brine recycle pump (flash only)
> effect pumps (multieffect only)

deaerated or decarbonated feed pump (acid-treated
  plants only)
distillate pump
condensate pump
air ejectors or vacuum pump, with inter- or after-
  condensing system

The duty of each pump on the actual plant, i.e., total head, flow rate, suction
conditions, the type of pump, nature of fluid to be handled, appropriate materials of construction and driver HP should be tabulated for comparison. Supplier diagrams and specifications are ideal teaching material.

Lecture 4: Seawater intake and outfall system:

size, arrangement, materials of construction of seawater
  intake pipe, trash racks, screens and supply piping to
  plant
chlorination and cathodic protection systems (if supplied)
cooling and brine discharge outfall system arrangement and
  materials of construction

Lecture 5: Steam supply, heat input and condensate return. Refer to Lecture
1 diagrams for orientation of the heater's function in the system and the steam
supply and condensate return.

steam supply piping and fittings, steam source, pressure,
  quality, temperature and flow available, description and
  function of pressure-reducing and desuperheating apparatus, main control valve and relief valve
heater mechanical design arrangements for both steam and
  brine flows; gas and air venting; design pressure; materials
  of construction
condensate drainage and return system

Lecture 6: Instruments and controls. Refer to specific plant flow diagram,
locate and identify all significant measuring points and their indicators or re-
corders where supplied, for:

temperature and pressure/vacuum
flow rate (indicating and integrating)
conductivity
pH
liquid level
watt-hr. consumption

Discuss importance of information given by each, and for the specific plant,
the normal settings and extremes allowable. Similarly, identify control locations
and describe their function in regulating:

steam supply pressure

heater outlet temperature
brine recycle flow rate
brine and distillate pump suction levels
heater hotwell level
acid injection rate

Lecture 7: Chemical treatment of seawater intake, makeup and distillate. Principles and practice of:

chlorination for marine growth control and hydrogen
  sulfide oxidation
additive treatments for scale control and antifoaming
acid treatment and decarbonation for scale control
pH control, disinfection and "hardening" of distillate
for storage, transmission and potability

Lecture 8: Plant startup, regulation, shutdown and layup procedures. The course should most fittingly conclude with specific practical instruction on full operating procedures, using the Instruction Manual supplied by the manufacturer.

CONCLUSION

This article was intended to be a stimulus to owner/operator management thinking, rather than an explicit manual on operator training. There are really no guidelines or precedents to follow for training because each facility in operation today possesses a unique combination of human, plant design and operational requirement factors. Those facilities which are regarded as efficient, safe and reliable, however, do have one common denominator, a competent and knowledgable staff. We hope that we have provided some perspective to help attain this goal.

THE REVERSE OSMOSIS MEMBRANE PROCESS

D.H. Furukawa
M.C. Morrison
D.R. Walker
Calgon-Havens Systems
Calgon Corporation

ABSTRACT

The reverse osmosis process using the preparation of membranes is discussed, and the different types of membrane structures described. The presentation concludes with discussion of the various concepts utilizing membranes for desalting systems.

INTRODUCTION

Reverse osmosis is a general term applied to a membrane process for separation and concentration of substances in solution. It usually consists of allowing a solvent to preferentially pass through a membrane normally described as semipermeable. The product is thus enriched in solvent, leaving a concentrated solution on the opposite side of the membrane. This process takes place at ambient temperatures and, thus, no phase change is involved.

Ultrafiltration and hyperfiltration are other terms used synonymously with this process. Differences, though subtle, do exist between ultrafiltration and reverse osmosis.

The term "osmosis" is generally described as the flow of solvent through a semipermeable membrane from a less concentrated solution to an area of higher concentration. A good example is the separation of water from saline water through an appropriate membrane. The normal osmotic flow of solvent can be reversed on application of hydraulic pressure to the more concentrated solution if such pressure is in excess of the normal osmotic pressure. Hence, the derivation of the term, reverse osmosis. Although this term is now universally used to describe this process, one would be more theoretically correct to say that it describes the preferential transport of material through the membrane in the direction of lower chemical potential under isothermal conditions. This thermodynamic requirement does not specify which component will be preferentially transported through the membrane.

The mechanism of the transport through the membrane and other theoretical aspects of the reverse osmosis process are still the topics of considerable debate. Reverse osmosis is used here as an acceptable description for the process rather than an accurate theoretical representation.

The mechanism which represents the most reasonable rationale for the process is that which was proposed by Dr. S. Sourirajan, who has been intimately involved in membrane technology for many years. It is hypothesized that the membrane effecting separation consists of a coarse film covered by a very thin active layer (1). The active layer contains a multitude of micropores and the separation of solute from solvent is the result of "molecular sieving."

By establishing this basic structure and mechanism, it is possible to explain the selectivities of various types of noncellulosic membranes. With an asymmetric structure and an active layer at the membrane-air interface, the sieving mechanism probably applies. That is, molecules are rejected by or transmitted through a membrane depending on the average pore size and the solute/solvent molecular weights and geometries.

Cellulose acetate membranes may transport solvents by a somewhat different phenomenon because of their ability to reject salt and still exhibit economic solvent transport. For cellulose acetate membranes, the mechanism for salt rejection is different from that for the rejection of organic solutes. Due to the peculiar physico-chemical nature of the cellulose acetate membrane surface, there is preferential adsorption of water to the membrane surface and repulsion of salt. In general, the repulsion increases with the valence of the salt. The result is a layer of pure water next to the membrane surface extending a thickness of about 10 angstroms (A.). This thickness represents two monomolecular layers of water.

If the pore size in the active layer is approximately twice the thickness of the preferentially adsorbed pure water layer, that is approximately 20 A., then ideally only pure water will pass through these pores under pressure.

The rejection of organic molecules is based on a sieve mechanism related to the size and shape of the molecule. Organic molecules are not repelled from the surface of the membrane. Because organics tend to lower the interfacial tension between the solution and the membrane, low molecular weight organics are enriched at the membrane surface. The pore size of a membrane rejecting greater than 97% sodium chloride is approximately 20 A. This membrane will reject all organics with a molecular weight greater than 200 and some with the molecular weights between 100 and 200, depending on the geometry of the molecule. However, this membrane will transmit a large fraction of organic molecules with molecular weights less than 100, such as phenol. There are, of course, anomalies which do not fit the pattern, such as organic compounds which act like salts (urea).

MEMBRANE PREPARATION

The coating solutions used to prepare desalting membranes contain three or more chemical compounds (2). First, there is a membrane polymer. At present, most commercially-manufactured reverse osmosis membranes are based on cellulose diacetate or a blend of cellulose triacetate and diacetate. Second, a solvent is required to place the polymer in solution, and third, a "modifying agent" is added. The last modifier is sometimes referred to as the "pore regulating agent" or "swelling agent" (3). It controls the formation of the skin and its microporous structure.

There has been extensive development toward a noncellulosic membrane. Noncellulosic polymer solutions generally produce membranes with greater pH resistance, retain active properties on drying and rewetting, and withstand larger temperature extremes.

There are two and sometimes three definite stages in the membrane production process. First, the polymer admixture is coated onto a porous fiber glass support tube or other suitable substrate. Second, depending on the nature of the coating solution and the substrate employed, an evaporation period may be required after coating and prior to the next stage. Third, the coated film is immersed into a nonsolvent, usually water.

During the immersion (third) step, the polymer film gels and membrane formation takes place. The membrane, consisting now of a polymer matrix and water (or nonsolvent) is heated to various temperatures prior to use. Membranes with widely differing selectivities are produced which can be used for applications ranging from desalting to fine screening of process streams or waste waters.

MEMBRANE STRUCTURE

The following discussion of the factors relevant to membrane structure will be limited to the cellulose diacetate polymer (average acetyl content 39.8%). There are three basic structures which have been observed in cellulose diacetate membranes.

Homogeneous and Dense

Consider the entire membrane thickness as being made up of dense polymer layers with a microporous network traversing the entire thickness of the membrane ($\sim$100μ or 4 mils) (4). These membranes might contain 30 to 40% by weight of the polymer and, although the micropores exhibit rejection characteristics, they generally have ,ow flux rates due to the large hydraulic resistance of the pore network.

Heterogeneous "Skinned" Membranes

These membranes are asymmetric in structure and exhibit two distinct regions. One is a dense layer about 0.2 to 1.0μ thick, containing micropores, and is referred to commonly as the "active layer" or active surface. This can be found at the air-membrane interface or the coating support surface-membrane interface. The remaining 99+% of the membrane thickness is composed of a much less dense polymer region, a matrix of macropores commonly called the substrate and approximately 1000 to 2000 A. in thickness.

The hydraulic resistance of the micropores and, basically, of the overall membrane is now approximately 1/100 that of the dense membrane, so the solvent transport is correspondingly higher. The rejection characteristics, however, are unchanged. This is the desired membrane structure ("skin" at the air-membrane interface) for all major process applications.

Homogeneous Porous

This membrane structure can be visualized as having uniform layers with low polymer density and a resulting overall macroporous structure. There is no active surface and, thus, even at low pressure, the solvent flux is correspondingly very good. The macropores, however, exposed to the process stream, suffer from boundary layer effects and fouling which, in turn, produce poor overall performance.

It has been hypothesized that the density of each layer in the membrane is determined by the concentration of the corresponding layer of coated polymer solution at the precipitation point (5). The volume concentration of the polymer at the precipitation point has been correlated to:

(1)   the solvation power of the solvent, as expressed by the amount of water (or nonsolvent) required for precipitation;

(2)   the direction of magnitude of osmotic volume flows taking place during the leaching (immersion) phase.

Table 1 shows the effect of various solvents on the structure of cellulose diacetate membranes. The various membrane structures can be related to the solvation power of the solvent and the osmotic volume flows during leaching. For acetone, the volume flow is entirely out of the polymer solution into the water (outflux). The polymer solution concentrates and, because there is no water flow into the solution (influx), there is no precipitation. A concentration is reached at which there is insufficient solvent to keep the polymer in solution and it precipitates, producing very dense layers throughout the membrane thickness.

TABLE 1: CELLULOSE ACETATE POLYMER

Acetyl Content = 39.8%

Solvent	Approx. Water Concentration at Precipitation Point	Membrane Structure	Solvent-Nonsolvent Exchange at Leaching		
			Direction of Osmotic Volume Flow	Water Influx	Water Outflux
Acetone	28	homogeneous very dense	solution to water	0.04	0.16
Dioxane	30	homogeneous dense	solution to water	0.08	0.14
Acetic acid	30	heterogeneous skin – airside	not measurable	-	-
Formic acid	-	heterogenous skin – airside	-	-	-
Dimethylformamide	15	heterogeneous skin – supportside	water to solution	0.16	0.08
Triethylphosphate	10	homogeneous porous	not measurable	0.07	0.07
Acetone plus (a) Formamide (b) Various inorganic salts	-	heterogeneous skin – airside	-	-	-

TABLE 2: CELLULOSE ACETATE E-398-3-FLAT SHEET MEMBRANE

Test Conditions: 600 psi, 5,000 ppm NaCl
Membrane Unheated Unless Specified

	Airside of Membrane		Supportside of Membrane		Water Content
	Flux, gsfd	% Rejection	Flux, gsfd	% Rejection	
Acetone	~0.1	>80	--	--	40
Dioxane	~0.4	>90	~0.4	~90	70
Acetic acid	6.0	>80	--	--	73
Formic acid	3.0	>95	--	--	70
Dimethylformamide	~270	None	25	80	78
Triethylphosphate	~110	None	100	None	75
Acetone plus:					
(a) Formamide cooked 82°C.	~20	~95	--	--	~64 - 70
(b) Inorganic salts cooked 70°C.	~20 - 30	95 - 90	~40 - 80	10 - 20	~64 - 80

With dioxane, the direction of osmotic volume flow is also from the polymer solution to the water; however, there is also considerable net water flow into the solution. The composition of the mixed solvent and water changes continuously during leaching. This difference between acetone and dioxane cast membranes is shown in their transport properties (Table 2). The dioxane membrane is less dense than its acetone counterpart.

With triethylphosphate (TEP) as a solvent, the solvent outflux during leaching is approximately equivalent to the water influx. Because only 10% water is required for precipitation, a homogeneous porous membrane is formed.

Consider now the formation of skinned membranes. The acetic acid/cellulose acetate polymer solution shows no noticeable net osmotic volume flux during leaching, with the solvent outflux and water influx approximately the same. However, as seen from Table 1, a high water concentration is needed for precipitation. During leaching, no concentration of the polymer can occur from osmotic volume flows and so, when the appropriate water concentration is reached, the solution precipitates from the polymer concentration existing in the coating solution. This produces a less dense substructure. The formation of the skin at the surface may be partly due to evaporation, or more likely, a direct result of the interfacial conditions existing at the precipitation point.

The situation arising with a cellulose acetate/dimethylformamide (DMF) is interesting. The direction of volume flow is from the water into the solution and a relatively small amount of water is required for precipitation. Hence, a porous, upper structure is formed. Under this condition, no dense layer can form at the water/coating solution interface. It is possible that the DMF flows out of the bottom layer of the coating solution and the polymer solution concentrates in that region prior to precipitation.

Reviewing these characteristics, acetone results in no water influx to the polymer and a very dense homogeneous structure is formed. Therefore, a membrane with extremely low flux, about 0.1 gsfd, is formed.

Using dioxane as the solvent, there is a considerable net water flow into the solution resulting in a less dense but still homogeneous membrane. By substituting dioxane for acetone, the flux is increased by 400% but is still quite low at 0.4 gsfd.

Using triethylphosphate, the water influx is approximately equal to the solvent outflux and this results in a homogeneous, very porous structure with a tremendous flux rate of approximately 110 gsfd. Unfortunately, the membrane has no rejection of NaCl.

By using acetic acid, a heterogeneous membrane with a skin and substrate is formed. Again, the water influx and solvent outflux are approximately equal but the skin formed exhibits a good rejection of sodium chloride and a reasonable flux of 6 gsfd.

From these observations it is obvious that with a given polymer, in this case cellulose acetate, a variety of membranes can be produced by changing the solvent system used for the coating solution.

"Skinned" cellulose acetate membranes can be produced from coating solutions using acetic or formic acid. It has also been shown that cellulose acetate/acetone (or dioxane) solutions, modified by suitable "pore regulating" or "swelling agents," can also produce heterogeneous membranes. The addition of a swelling agent alters the osmotic volume flow pattern at leaching and thereby produces an asymmetric, skinned membrane.

SELECTIVITY AND PORE SIZE

For a given cellulose acetate content, it is possible to alter the swelling agent to solvent ratio in the coating solution and produce membranes with differing selectivities. An increase in this ratio can produce a more "open" membrane, i.e., one with a larger pore size, allowing the passage of higher molecular weight solutes. The assumption is made that "open" membranes with larger pores correspond to membranes with less dense active layers.

The pore size and the number of pores in the dense, active layer of a heterogeneous membrane can be discussed in terms of polymer-solvent interaction. If at the precipitation point, the distance between polymer chains is at a minimum, then the active layer formed will be dense with average size micropores. This situation might arise in a coating solution where the polymer-solvent interaction was low.

Increasing the swelling agent content would affect the polymer-"modified solvent" interaction. This is possibly now greater, resulting in a larger interchain distance and, at the precipitation point, an active layer is formed which although dense, has a larger average pore size.

For cellulose acetates it has been observed that the effect of increasing the swelling agent concentration (formamide) differs with the solvent system employed. This phenomenon is probably due to the fact that the polymer-"modified solvent" interaction will be dependent on the nature of the solvent for a given swelling agent.

MEMBRANE SUPPORTING STRUCTURES

Having considered the mechanisms for the reverse osmosis process, membrane materials and manufacturing technique, one must consider an adequate support on which to place this very thin membrane. Generally, a rigid porous support is provided which will withstand the normal operating pressure required. The most elementary, of course, would be a porous stainless steel disc. The available designs for reverse osmosis equipment can be categorized into the plate and frame (6), spiral wound, tubular and hollow fibers (7).

The plate and frame design was used in the earliest reverse osmosis systems. Early engineers in the field, seeing the great potential of membrane separation processes as a unit operation, no doubt noticed the similarities between the new process and another unit operation, filtration, and constructed a reverse osmosis system similar to a plate and frame filter press. The membrane is supported on a porous, flat support plate. By alternating membranes and their support plates with channel spacers, and by arranging suitable manifold ports for the feed and purified waters, large volume systems can be built. Several disadvantages have been found in the past with plate and frame units:

First, membranes must be handled extensively during unit assembly and disassembly. Because of the weak nature of membranes, this extensive handling is conducive to failure. Second, flow distribution arose because of the use of small channel heights to optimize the surface area/unit volume ratio. Third, because of the labor costs involved, systems using this concept are expensive to maintain when relatively few system failures occur.

The only company known to be using this concept now for desalination purposes is a firm in Denmark. For low pressure ultrafiltration units, Dorr-Oliver has developed a plate and frame module which purportedly overcomes disadvantages noted. This is an arrangement in which the membrane is applied to a porous, replaceable cartridge which is inserted into a molded rectangular shell and cover piece. Feedwater flows over the membranes; purified water flows through the porous support and out of the module.

The spiral wound concept is a popular one in the marketplace today. Two membrane sheets are separated and supported by a porous backing material and are glued together at the ends. In the final module, these alternating mesh spacers and membranes with their porous supports are spirally wrapped about a porous central tube which serves as a collecting duct for the permeate. The feed enters the unit axially. Product water is collected in the porous supports and flows through the support to the porous central collecting tube. Product water leaves the module through this tube.

This concept has a higher membrane surface area/unit volume ratio than the tubular concept. It also requires a less costly high pressure support vessel per unit of membrane area than the tubular design. A primary disadvantage when compared with tubular units, is encountered in handling feed streams when moderate solids are present in the feed or are precipitated as water is removed.

The tubular concept for design of reverse osmosis equipment such as that produced by Calgon-Havens Systems overcomes the primary disadvantage of the spiral wound concept, but does have its own disadvantages. In the tubular system, the feed stream enters through the center of the tube. The membrane is coated on the interior tube wall. Purified water passes through the membrane, through the porous tube wall, and is collected outside the tube. Obviously, such a design is not limited by the presence or formation of moderate solids. However, it does have a lower membrane surface area/unit volume ratio, and it requires a relatively expensive support structure per unit area of membrane. One way to lower the cost of the tubular support is to take advantage of the relatively higher strength of a structure in compression than in tension. This is done by placing the membrane on the tube exterior. The feed passes over the tube exterior and purified water is removed through the tube interior. Materials used in tube construction include fiber glass/resin composites, stainless steel, bronze, aluminum, porous ceramic and sand/resin composites.

One of the primary disadvantages of a tubular system is its low membrane surface area/unit volume ratio. This ratio is increased as one decreases the diameter of the tube. This observation gave birth to the hollow fine fiber concept. In this module, the membrane material is spun into hollow fibers with a very fine diameter, usually 1 to 5 mils. By compressing a large number of these fibers into a cylindrical bundle and potting the ends of the fibers in a plug of resin, a module can be fabricated with an unusually high membrane surface area/unit volume ratio.

The bundle of fibers is inserted into a high pressure cylinder to serve as the containing vessel. Feed enters the end of the module and flows over the outside of the hollow fine fibers. The purified water passes through the fiber membrane wall and is removed through the interior of the fiber. Note that, in this configuration, because of the very small dimensions involved, the compressive strength of the membrane material is sufficient to allow operation of the module at relatively high pressure without rupturing the unsupported membrane material.

Hollow fine fibers, with their very high membrane surface area/unit volume ratio, enable the construction of very compact systems. To date, solvent transport achievable with hollow fine fiber membranes is more than one order of magnitude less than membranes cast in flat sheet or tubular form, but this does not negate the advantage of the substantially greater membrane surface area/unit volume. The hollow fine fiber modules possess the disadvantage that they are quite susceptible to plugging by suspended solids, laden feed streams and by fouling during operation.

Each of the four membrane configurations discussed — plate and frame, tubular, spiral wound and hollow fine fiber — has its relative advantages and disadvantages. At the current time, no one configuration is able to prove superiority in every situation. Which configuration is used in a particular system depends on the characteristics of the feed stream to the system and the desired results from the system.

SYSTEMS DESIGN

There are several system designs possible which will maintain critical velocities in a reverse osmosis system (8),(9),(10),(11),(12). The most desirable system is a once-through design. It is essential in reverse osmosis system design to maintain critical minimum velocities to overcome the phenomenon generally referred to as concentration polarization (13),(14). In order to maintain critical velocities and adequate pressure, however, a once-through system must be designed in stages with booster pumps between stages. The disadvantages to this system are that large holding tanks are required and long detention times are involved. The principal advantage is that a fewer number of modules is required, reducing capital cost.

For small systems the batch or internal recycle design can be utilized. In the internal recycle design, a fraction of the concentrate is recycled directly back into the feed stream. The remaining fraction is "bled-off" at the desired concentration. Again, the advantage is that a smaller number of modules are required, but the disadvantage of long detention time is a factor.

Systems for production of large volumes of potable water require significantly more sophisticated designs. Full-scale commercial systems also require more complete systems control and instrumentation. Multimillion gallons per day plants may necessitate hybrid computer control.

In the last ten years, reverse osmosis has developed from a laboratory curiosity to a viable commercial unit operation. A critical development period is immediately ahead as larger systems are built. As experienced with other desalting processes, new problems will develop as the size of systems increases. Reverse osmosis has already established its importance as a separations process in chemical engineering. The future promises bigger and better desalting plants through use of this significant process.

REFERENCES

(1)   Sourirajan, S., Reverse Osmosis. Academic Press, 1970, New York.

(2)   Merten, U., Desalination by Reverse Osmosis. M.I.T. Press, 1966, Cambridge, Massachusetts.

(3)   Manjikian, S., "Desalination Membranes from Organic Coating Solutions." I&EC Product Research and Development, Volume 6 (March, 1967), pp. 23-32.

(4)   Merten, U., "Reverse Osmosis." International Desalination Symposium.

(5)   "The Mechanism for Formation of a 'Skinned' Type Membrane." Final Technical Report, Hydronautics, Inc., Israel, O.S.W. Contract No. 14-01-0001-1706.

(6) Keilin, B., "A Survey of Desalination by Reverse Osmosis," ASME Publication 67-UNT-7, ASME United Engineering Center, New York.

(7) Furukawa, D.H., "RO: Packed with Potential." Water & Wastes Engineering, November, 1971.

(8) Cooke, W.P., "Permasep Permeators in Industrial Waste Stream Separations." Effluent & Water Treatment Journal, February, 1970.

(9) Keilin, B. and DeHaven, C.G., Design Criteria for Reverse Osmosis Desalination Plants.

(10) Kimura, S., Sourirajan, S., Ohya, H., "Stagewise Reverse Osmosis Process Design." I&EC Process Design & Development, Volume 8, No. 1 (January, 1969).

(11) Srinivasan, S. and Chi Tien, "Reverse Osmosis Desalination in Tubular Membrane Duct." 2nd European Symposium on Fresh Water from the Sea, Athens, Greece, May 9-12, 1967.

(12) Merten, U., "Flow Relationships in Reverse Osmosis." I&EC Fundamentals, Volume 2, No. 3 (August, 1963).

(13) Kimura, S. and Sourirajan, S., "Concentration Polarization Effects in Reverse Osmosis Using Porous Cellulose Acetate Membranes." I&EC Process Design & Development, Volume 7, No. 1 (January, 1968).

(14) Sherwood, T.K., Brian, P.L.J. and Fisher, R.E., "Salt Concentration at Phase Boundaries in Desalination by Reverse Osmosis." I&EC Fundamentals, Volume 4, No. 2 (May, 1965).

DESALINATION SYSTEMS BASED ON REVERSE OSMOSIS

Gerald M. Whitman and Vincent P. Caracciolo
Organic Chemicals Department
E.I. du Pont de Nemours & Co.
Wilmington, Delaware

ABSTRACT

Typical systems for carrying out reverse osmosis operations are described with emphasis on results with hollow fiber membranes. Description of the operation of B-9 Permasep permeators is followed by discussion of complete reverse osmosis systems, including operating parameters, pretreatment and posttreatment of water, and design of municipal water systems. Costs of desalting brackish water are considered in general and in relation to two specific installations for supplying potable water. Several other types of reverse osmosis applications are described, with emphasis on treatment of industrial water.

HOLLOW FIBER MEMBRANES

My portion of this discussion of reverse osmosis is built around the Du Pont Company's particular contribution to reverse osmosis technology – namely, the development of special membranes in hollow fiber form and the development of special devices for using hollow fibers to desalt water.

I will first discuss typical systems for carrying out reverse osmosis operations, and even though I discuss them in terms of hollow fiber membranes – because that is where our experience lies – much of what I have to say will apply to other forms of membranes such as the tubular form, flat films and spiral-wound membranes. I will then describe two specific examples of the use of permeators to supply potable water including a breakdown of the costs involved, and finally describe some of the applications of reverse osmosis other than desalination.

Du Pont has spent more than ten years and a large amount of technical effort to find the most efficient method of desalting brackish waters. The result of this work is the B-9 Permasep permeator. The B-9 permeator represents a major step in the commercialization of reverse osmosis through use of an asymmetric hollow fiber made from an aromatic polyamide. Figure 1 shows the cross section of a B-9 fiber.

FIGURE 1: CROSS SECTION OF HOLLOW FIBER

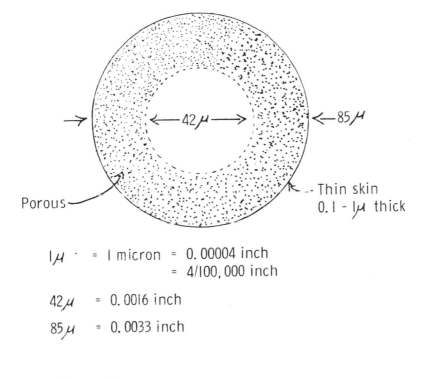

1μ = 1 micron = 0.00004 inch
= 4/100,000 inch

42μ = 0.0016 inch

85μ = 0.0033 inch

Asymmetric

Aromatic Polyamide

The hollow fibers are self-supporting membranes and are actually thick-walled cylinders. The ratio of outside to inside diameter is 2 to 1 (namely 84 to 42 microns) and this gives the fibers the strength to withstand high external pressures without collapsing.

The polyamide membrane has outstanding flux and salt passage characteristics and a wide range of resistance to both chemical and biological attack. We feel it represents a major improvement over cellulosic-based membranes. The use of hollow fibers permitted the development of an extremely compact device because their small diameter, about the size of a human hair, results in a large surface area for water transport. For example, the comparative surface areas of membrane available in a cubic foot of space in various devices are as follows:

(a) Tubular device   100 sq. ft./cu. ft.
(b) Spiral wound device   300 sq. ft./cu. ft.
(c) Hollow fiber device (B-9)   5,000 sq. ft./cu. ft.

The advantage of the hollow fiber device in this respect is obvious.

Du Pont officially introduced the B-9 permeator in December 1970. The initial size offered was 4 inches in diameter and 4 feet long. This size permeator contains almost a million hollow fibers. This permeator, when operated at 400 pounds per square inch, 20°C., and 75% conversion will produce over 2,000 gallons of product water per day containing less than 10% of the salts originally present in the feed water, which contained 1,500 ppm NaCl.

Figure 2 shows more fully just how the permeator works. Essentially the permeator is a 4 inch i.d. pipe filled with a bundle of hollow fibers each of which is in the form of a loop, with the open ends encased in a casting of epoxy resin. The epoxy forms a seal which separates the feed water on one side from the purified product water on the other.

FIGURE 2: CUT-AWAY DRAWING OF PERMASEP PERMEATOR

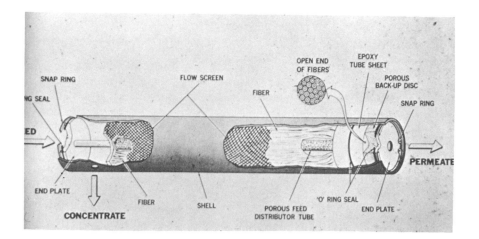

The feed water at 400 psi pressure enters the permeator through a very porous distributor tube located at the center of the permeator. This distributor runs the entire length of the permeator. The feed water moves radially from the distributor toward the outer shell of the permeator and around the outside of the fibers.

The 400-pound pressure forces essentially pure water through the fiber walls into the bore of the fibers, and this water moves along the bore of each fiber to this end of the permeator where the fibers have been cut to allow the pure water to escape and be collected.

The salts remaining in the reject water move to the outer perimeter and are taken out of the permeator through this reject port. Going back now to the epoxy resin casting, once the epoxy resin has cured, we put the bundle in a lathe and with a very sharp knife cut across the face of the epoxy casting to open the ends of fiber. Figure 3 shows a photomicrograph of the open ends of the fibers. We are successful in opening almost 100% of the fibers without distortion.

FIGURE 3: PHOTOMICROGRAPH OF OPEN ENDS OF HOLLOW FIBERS

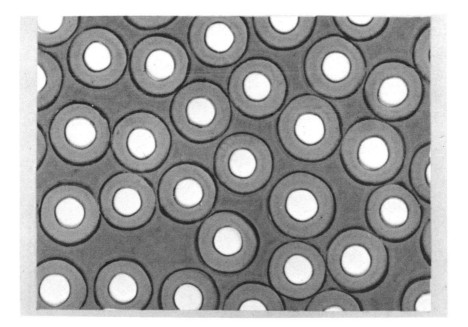

I mentioned earlier that our initial product was 4 inches in diameter and 4 feet long. We have just introduced a product which is 8 inches in diameter and 4 feet long. This permeator will produce 7,500 gallons per day of product. It weighs only 150 pounds, so heavy equipment to handle these permeators is not necessary. Figure 4 illustrates 8 and 4 inch permeators.

FIGURE 4: 8-INCH AND 4-INCH PERMASEP PERMEATORS

Operation at 400 psig pressure is an optimized condition which minimizes power consumption for pumping and maximizes output or flux at the end of three years, which is our guarantee period. The permeator can operate over a wide range of conversions from 25 to 90%.

For systems requiring larger capacity, individual permeators can be assembled into modules, as shown in Figure 5. Each individual permeator operates just like every other one under the same conditions, so it is possible to build a very large plant simply by placing permeators in parallel.

In this module, the feed water, the product water, and the reject stream each are piped through common headers. When permeators are operated in parallel in this way, an additional 50 psi pressure drop is built into each permeator reject stream by means of an orifice or length of small high pressure tubing. This additional pressure drop balances the water flow among the permeators.

FIGURE 5: MODULE CONTAINING TWELVE 4-INCH PERMEATORS

This modular unit of 12 4-inch permeators occupies only 5' by 5' by 3.5' and has a capacity of 24,000 gpd. These modules can be moved and arranged as units. The permeators are connected in a true parallel fashion, so that testing individual permeators or adding a permeator is quite easy. Changing a permeator requires only a few minutes.

THE BASIC REVERSE OSMOSIS SYSTEM

So far we've been discussing only the permeators. Obviously a complete reverse osmosis system requires much more than just the permeators themselves. All the components of a basic reverse osmosis system are illustrated in Figure 6, which shows a schematic drawing of a complete one million gallon per day plant.

The components normally required are pretreatment equipment such as cartridge filters and chemical feeders; high pressure pumps; reverse osmosis modules; and posttreatment equipment, such as a decarbonator, surge tank and chemical feeders.

FIGURE 6: SKETCH OF ONE MILLION GALLONS PER DAY PLANT

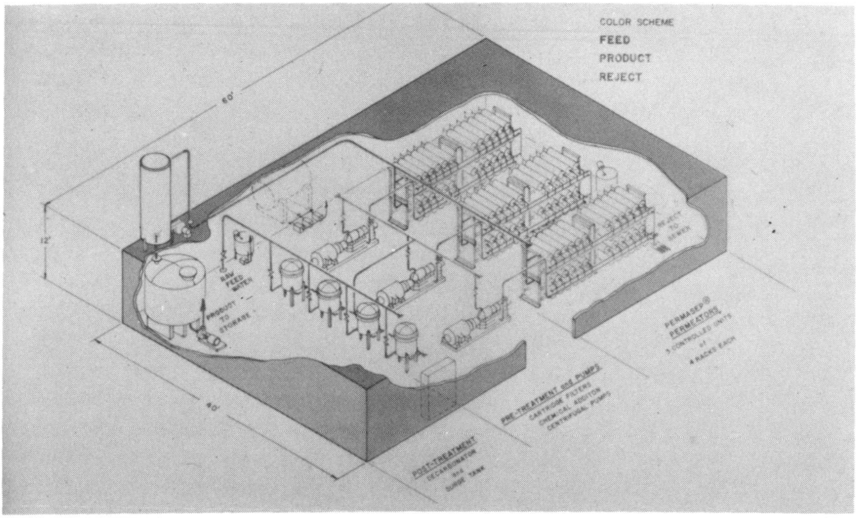

Feed water is pressured by pumps servicing each block of reverse osmosis modules. Minimum filtration of feed water is always provided by cartridge filters (5 to 10 micron pores) as protection for pumps and permeators. Pretreatment with conditioning chemicals may or may not be required, depending on the chemical constituents in the feed water. Likewise, product water may or may not be posttreated before use. Space requirements for this typical one million gallon per day plant are 40 by 60 feet or a total of only 2,400 square feet.

Now a few other details about operation. Control over the operation of a reverse osmosis plant is quite straightforward. A properly designed plant can run almost unattended. Plant output is controlled by feed water pressure. Conversion is regulated by adjusted reject flow to the proper ratio relative to product flow. Municipal plants are generally designed to operate in the on-off mode, controlled by demands from the water storage units that are part of the municipal distribution system.

Operating reliability as high as 97% on-stream availability has been demonstrated. Only a minimum of operating labor and maintenance is required. For example, requirements for monitoring instruments and making minor plant adjustments in a typical one million gallon per day plant are normally less than one hour per day. Maintenance on pumps and on chemical and other supplies averages less than one hour per day.

FIGURE 7: 24,000 GALLONS PER DAY REVERSE OSMOSIS PLANT

Figure 7 shows a small 24,000 gpd plant. This facility contains all auxiliary equipment necessary for plant operation – chemical feed, pumps and control instrumentation. This complete small plant occupies a space of 5' by 7' by 4½'.

Obviously, arrangements of permeator modules other than the completely parallel arrangement are possible. For example, a multipermeator system could consist of a number of permeators operating in parallel, but in two stages instead of only one.

In such a two-stage system the second stage could process either the reject stream from the first stage permeators or the product stream. Operation where the reject from the first stage becomes the feed for the second stage is an advantage when very high conversions are desired or the water contains precipitating salts or colloidal material.

In the latter case, conversion is kept low in the modules by maintaining a substantial reject flow rate. This tends to keep potential foulant materials moving through the bundle and makes it easier to handle waters which tend to precipitate solids. A typical arrangement of this is shown in Figure 8. The reject streams from the first stage become the feed for the second stage.

In cases where the level of dissolved solids is higher than 5,000 ppm, water meeting public health standards of less than 500 ppm total dissolved solids

FIGURE 8: SERIES OPERATION — REJECT STAGING

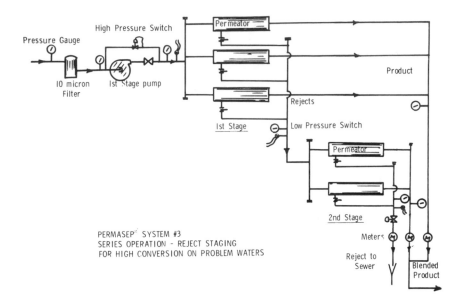

cannot be met in a single stage at reasonable conversion levels. In this case the product from the first stage is repressurized to 400 psi and processed through a second stage.

This system is shown in Figure 9. In most instances the product from the second stage will then contain a very low solids level and can be blended with product water from the first stage to yield water within Public Health Service standards at decreased overall cost.

In addition, the reject from the second stage contains less solids than the first stage feed and is, therefore, blended with the first stage feed. The result is good quality water at good recovery and minimum cost.

FIGURE 9: SERIES OPERATION — PRODUCT STAGING

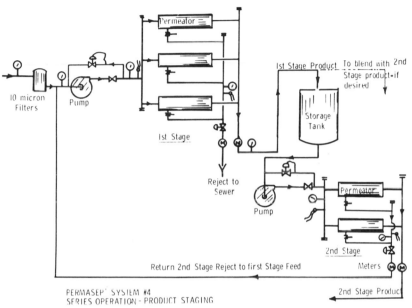

PERMASEP SYSTEM #4
SERIES OPERATION - PRODUCT STAGING
FOR HIGHLY SALINE WATERS

Materials of Construction

In general, all high pressure piping should be Type 316 stainless steel for brackish water service. Carbon steel is not practical because most systems operate at a pH slightly below 6 to control scaling. Low pressure piping (pump suction, product piping, and reject piping downstream of the flow control valve) should be polyvinyl chloride (PVC) or other plastic. PVC has been successfully used for 400 psi service but is limited to small diameters.

Aluminum was originally used exclusively for the permeator shells and found satisfactory for most brackish water uses. This was Type 6063-T6 alloy. However, filament-wound fiberglass shells are now being used for both the 4-inch and 8-inch size permeators. The fiberglass shells have been shown to be very corrosion-resistant and have permitted the use of B-9 permeators over their full range of pH resistance, which is from pH 4 to 11.

Pumps with bronze, stainless, and plastic components have all proved satisfactory.

Application in Potable Water Systems

Figure 10 presents a sketch of a typical application of reverse osmosis to a municipal water system. Normally discussion of system design centers around the basic reverse osmosis process and ignores other design considerations specific to the particular use situation, such as water source, feed water pumping, quality of desired product, and waste disposal.

Such considerations are important for the design and costing of a complete municipal system because they determine requirements for special pretreatment, allowable plant conversion, and allowable blending. I can't stress enough that every specific situation has to be analyzed on its own merits, and generalizations are dangerous.

The source can be either ground water or surface water. By nature, ground waters are more suitable because of their lower levels of turbidity. Special pretreatment of surface waters is required to remove excess turbidity. Basic cartridge filtration is required in any event. Depending on the chemical analysis of the feed water, other special pretreatment outside of the basic reverse osmosis process may be required.

FIGURE 10:   TYPICAL MUNICIPAL SYSTEM

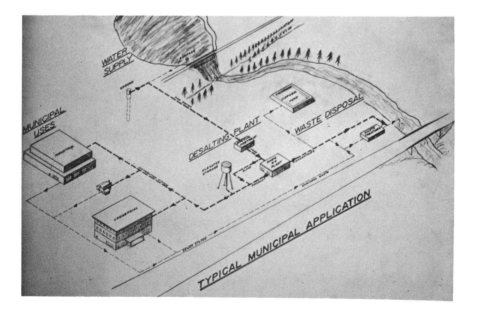

There is considerable leeway in the conversion level at which a plant is designed to operate. Use of preconditioning chemicals allows a reverse osmosis plant to be operated at higher conversion, thus saving feed water pumping costs and waste disposal costs.

Assuming that there is an adequate supply of feed water, an economic balance must be struck between the cost of pretreatment chemicals on the one hand and the cost of feed water pumping and waste disposal on the other.

The volume and concentration of waste for disposal is determined by the level of conversion chosen. In a typical one million gallon per day plant operated on a 2,000 ppm feed water at 75% conversion, the volume of waste is 333,000 gallons/day at a concentration of about 8,000 ppm.

Since reverse osmosis normally produces a product water better than that required by health standards, blending of product with the feed water offers a logical and practical way of reducing water treatment costs. Both plant investment and operating costs can be reduced in direct proportion to the amount of blending.

For example, about 45% of a feed water with 1,000 ppm dissolved solids can be blended with a 100 ppm product to make a 500 ppm water for delivery which meets Public Health Service standards with a proportionate reduction in costs.

Costs

This brings me to the subject of costs, on which any system will eventually stand or fall. I'd like first to make some generalizations about costs in reverse osmosis systems and then go on to discuss some specific examples.

Desalting costs are usually expressed on a battery limits basis. This includes the basic desalting process but not the cost of acquiring feed water, special pretreatment, product water storage or waste disposal. As we've said, the latter costs can vary substantially depending on the specific application and should be considered fully when comparing reverse osmosis to conventional alternatives.

To review the ground rules, total operating costs for a battery limits reverse osmosis plant are divided into operating and amortization costs. Operating costs include the costs of direct operating labor and of consumables such as power, chemicals, and filters.

Cost of membrane replacement is based on membrane price and life. Amortization covers the cost of initial plant purchase and is based on the cost of money (interest) and the financing period. Normally, suppliers of desalting equipment will guarantee items such as total installed plant costs, the costs of consumables, and costs of membrane replacement.

Generalizations about reverse osmosis costs are often subject to misinterpretation because they depend critically on a dozen or more design and price assumptions that are specific for each application.

Particularly significant design parameters that affect cost are plant size, conversion level, load factor, and blending. But with these reservations in mind, I will generalize that capital costs for relatively small reverse osmosis plants (30,000 to 50,000 gpd range) can be as high as $1.00/gpd of capacity, while larger plants (3 to 5 million gallons per day) can be as low as $0.25/gpd of delivery capacity.

The total operating costs for these plants can range from over $0.65/1,000 gallons to below $0.25/1,000 gallons of delivered water. These are battery limits operating costs and do not include costs of feed water supply or waste disposal.

Practical Applications

Now I'd like to leave the generalizations and discuss costs in two specific installations using permeators to supply potable water. The first is a new trailer park near Sarasota, Florida, with some 250 trailer homes planned. Its well water is high in hardness and exceeds the Florida Health Service standards with respect to total dissolved solids and sulfate. Suitable supplies of good water were not available in the area and a pipeline from the nearest city water was too expensive.

A small 24,000 gpd plant using Permasep permeators was installed early in 1971, as can be seen in Figure 11. The building behind the permeators houses the high pressure pump and chemical feed systems. The product water can be blended with about ¼ its volume of feed water so that a total of 30,000 gpd can be delivered to the trailer park. The plant operates at 66% conversion and concentrate is disposed of through the park's sanitary sewage system. The plant is projected to operate at about a 90% load factor. Performance and costs of this system are given in Table 1.

TABLE 1: PERFORMANCE AND COSTS OF TRAILER PARK INSTALLATION

Water Quality

Mineral	Raw Feed	Permasep Product	Blended Water (Meeting USPHS Standards)
Calcium	312	10	73
Magnesium	141	12	40
Sodium and potassium	60	15	25
Bicarbonate	183	37	67
Chloride	96	59	67
Sulfate	1,008	21	228
Iron	0.05	Nil	0.01
Silica	24	4	8
Hardness	1,360	60	333
TDS	1,800	155	500

(continued)

TABLE 1: (continued)

Permasep Permeator Plant Costs

Installed Cost – $30,100

Projected Operating Costs		Cents/1,000 Gallons Blend
Power @ 1 cent/kwh		5.8
Chemicals		8.3
Filters		1.6
Labor @ $4/hr.		10.4
Membrane replacement		13.4
	Sub-Total	39.5
Amortization*		26.0
	Total	65.5

*Amortization at 6% interest over 20 years

FIGURE 11: 24,000 GALLONS PER DAY PLANT FOR TRAILER PARK

Product water better than Public Health Service standards is produced, with over 90% rejection of hardness and total dissolved solids. As can be seen by comparing the dissolved minerals in the product column to the raw feed, all minerals are substantially reduced.

Actual overall salt passage is less than 9%. A blend of water containing 500 ppm total dissolved solids is delivered. This 24,000 gpd plant cost $30,100 completely installed, without the building. Total power consumption is 7.3 kwh per 1,000 gallons of product water. Chemical costs include sulfuric acid and Calgon.

Membrane replacement is based on an expected life of five years. Operator attention required is about 5 hours per week. Total operating costs under these conditions, including amortization, are a little over 65 cents/1,000 gallons of delivered water. The operating costs are lower than might have been expected for this size plant because of the blending previously described.

The second installation is at Greenfield, Iowa, which has a population of 2,300. It is the world's largest municipally-owned reverse osmosis plant. The plant is a standby facility operated on a brackish well only during summer drought periods when Greenfield's peak demand cannot be satisfied by the normal surface water supply. The well exceeds the U.S. Public Health Service standards with respect to sulfate, fluoride, iron and total dissolved solids.

The Greenfield plant has a nominal capacity of 150,000 gpd, with room for expansion to 300,000 gpd, and produces water within recommended health limits. Figure 12 shows two 50,000 gpd banks of modules stacked on top of one another and one 50,000 gpd bank located across the aisleway in the town's conventional water treatment plant.

Pretreatment includes sulfuric acid and Calgon. Special pretreatment to remove the high concentrations of iron salts was considered, but was not adopted. Instead, periodic membrane cleaning is carried out. Posttreatment consists of decarbonation followed by blending with the finished surface water. The plant operates automatically at 67% conversion. Performance and costs of the Greenfield plant are shown in Table 2.

TABLE 2: PERFORMANCE AND COSTS OF GREENFIELD PLANT

Water Quality

Mineral	Raw Well Water	Permasep Product	Surface Water	Blended Water
Calcium	160	2	50	26
Magnesium	49	0.4	11	6
Sodium and potassium	410	41	13	27
Bicarbonate	247	59	154	106

(continued)

TABLE 2: (continued)

Mineral	Raw Well Water	Permasep Product	Surface	Blended Water
Chloride	351	32	8	20
Sulfate	864	29	32	30
Fluoride	2.1	1	0.4	0.7
Iron	2.9	0.05	0.05	0.05
Hardness	600	6	170	88
TDS	2,100	170	210	190

Permasep Permeator Plant Costs

Installed Cost — $94,300

Projected Operating Costs	Cents/1,000 Gallons Product
Power @ 1 cent/kwh	7.0
Chemicals	16.0
Filters	5.0
Labor – O & M	5.0
– Cleaning	3.0
Membrane replacement	16.0
Sub Total	52.0
Amortization*	16.0
Total	68.0

*Amortization at 6% interest over 20 years

FIGURE 12: 150,000 GALLONS PER DAY PLANT FOR GREENFIELD, IOWA

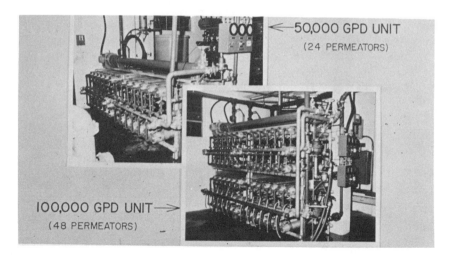

←50,000 GPD UNIT
(24 PERMEATORS)

100,000 GPD UNIT→
(48 PERMEATORS)

Well water total dissolved solids are reduced from 2,100 to 170 ppm. Over 98% of the hardness is rejected. Overall salt passage is 8% and the water pro- duced is better than the surface water normally used.

The Greenfield plant cost $94,300 completely installed within the existing building. Power consumption is 7 kwh/1,000 gallons of product water. Clean- ing frequency for iron removal is once every 3 to 4 months and takes 4 hours per bank of 24 permeators to complete.

Membrane replacement is based on a five-year expected life. Total operating costs are projected at 68 cents/1,000 gallons of product for 95% load factor operation. These costs are higher than might be expected for this size plant normally because of the high iron levels in the feed water and the consequent necessary interruptions for cleaning the membranes.

OTHER APPLICATIONS

Now finally a word about other applications of reverse osmosis. So far we have been discussing desalination of brackish waters not exceeding 5,000 ppm of total dissolved solids. Permeators have been and are being successfully used on water containing 20,000 ppm total dissolved solids producing product of less than 2,000 ppm in one stage.

The Grand Bahamas Hotel is a good example of this. A second stage is used to lower the total dissolved solids to potable levels. Although we are investigating seawater desalination, we do not feel that we have a commercial product to of- fer in this field at this time.

These permeators are very efficient in removing a wide variety of materials from water in addition to inorganic salts. First let me mention bacteria, viruses, and pyrogens. Tests performed by customers as well as by Du Pont show that essentially all of these organisms are removed. We cannot make claims for 100% removal because any device with a mechanical seal is subject to very minor leaks, and we can't guarantee sterile water.

Permasep permeators have the capability of separating and concentrating dis- solved materials from a wide variety of industrial wastes. There are three gen- eral incentives for doing this. One is the recovery of valuable components from a waste. A second is the recovery of water for reuse, especially in high cost water areas.

A third incentive, of course, is pollution control. The concentrated stream from Permasep treatment of an industrial waste can often be more easily and economi- cally disposed of by established means such as evaporation or burning, or it may be recycled to the process.

We have amassed considerable data on our separation capabilities for both inorganic and organic solutes, and are continuing to do so. These data include not only "in-house" results, but field trial work in actual industrial waste situations.

One illustration of waste treatment is treatment of rinse water from electroplating operations containing heavy metals. Typical field test data on nickelplating rinses and on chromium-plating rinses are shown in Table 3 for a B-9 permeator operating under standard conditions and approximately 75% conversion.

TABLE 3: CONCENTRATION OF NICKEL- AND CHROMIUM-CONTAINING RINSES

Plating Waste Separations (ppm)

	Feed	Product	Reject
Nickel as Ni	4,600	231	17,800
Chromium as Cr	42	1	154

With these separation capabilities, it is possible to design a completely "closed loop" type of system for electroplating lines, and a commercial unit has been designed for early installation in a nickel-plating rinse line. The system provides for recycle of permeate for reuse as rinse water, and recycle of concentrate to the plating bath itself. There will be no discharge of nickel from the process. Savings in chemicals that are now discarded to the sewer will pay for the installation in less than one year.

The removal of organic compounds has been investigated. With organic compounds, the molecular weight and the type of molecule are both significant. Table 4 shows typical removal of organics. You see that in many cases the removal is close to 100% per pass.

TABLE 4: ORGANIC SEPARATIONS

Separation/Concentration of Organic Materials

Compound	Molecular Weight	Concentration (ppm)	pH	Rejection (%)
Carbohydrates				
Raffinose	504	2,000	7	99
Sucrose	342	500-2,000	5-8	99
Sorbitol	182	2,000	7	99
Glucose	180	500-2,000	5-8	99
Arabinose	150	2,000	7	99
Glycerol	92	500-2,000	6-7	90
Ethylene glycol	62	2,000	6-7	48

(continued)

TABLE 4: (continued)

Compound	Molecular Weight	Concentration (ppm)	pH	Rejection (%)
Acids				
Phenol	94	500-2,000	7-9	55
Acetic acid	60	500	4	40
Sodium acetate	82	500	8	98
n-Butyric acid	88	500-2,000	3-4	70
Pivalic acid	102	500-2,000	4	98
Benzoic acid	122	500	4	83

The potential applications for reverse osmosis in the treatment of waste range from heavy metal-containing situations such as the one just described, to acid mine drainage, which represents a major source of pollution in certain areas of the U.S., and the demineralization of effluents from secondary waste treatment plants in both industrial and municipal installations. We look to the treatment of waste as a major area for sales of reverse osmosis equipment in the future.

Finally, I'd like to take a minute for a brief commercial. Some of you may already know, but others probably do not, that it was recently announced that Du Pont has been named the winner of the 1971 Kirkpatrick Chemical Engineering Achievement Award for the development of the B-9 permeator. This award honors group achievement by chemical engineers and is given biennially. We take considerable pride in having won this award, and I thought it would be appropriate to mention it to you.

THEORY OF DISTILLATION BASED OPERATIONS

Richard M. Ahlgren
Aqua-Chem, Inc.

ABSTRACT

The distillation based approaches are reviewed considering the theoretical factors involved. The applications of these principles in commercial desalination operations are presented.

DISTILLATION BASED DESALINATION OPERATIONS

Distillation type desalination operations all involve phase change or the boiling of saline water. Phase change is the conversion of the physical form of a specific chemical (water) from the liquid state into a different state such as vapor. Obviously, to make the process useful, this phase change must be reversible from saline water to pure water vapor then back to pure liquid water.

There are two general modes of operation for the equipment to carry out such evaporation operations. The broad division between these two categories is whether the boiling or phase change operation takes place on a discrete heated surface or whether the phase change takes place from a pool or body of water.

In order for any phase change to vapor state to occur the liquid saline water must have an adequate amount of energy added to it from outside sources. The general source of energy can be either fresh heat from steam boilers or electrical sources or recycled or regenerant energy from within the system itself.

Evaporation equipment for desalination can be divided into broad categories by the presence or absence of a discrete heat transfer surface. Those types of equipment operating with a definite heat transfer surface on which boiling or phase change occurs are the evaporation forms such as submerged tube, vertical tube, thin film, wiped film, spray film, forced or natural circulation, and several others. Thermodynamically, these can be coupled in several different modes and can operate as either single or multiple effect steam motivated units or as in the vapor compression cycle. Mechanical or jet type compressors can be used for the latter mode.

The evaporator types operating without the use of a discrete heat transfer surface are typically the flash and solar distillation approaches. In the flash mode they can be operated as either single or multiple effect and can function in either the once through or recirculation type plants.

Many variations on this scheme and combinations with other types of plants has been practiced in recent years. The most common form of solar distillation equipment utilizes the direct boiling of a pool of saline water with subsequent condensation of the pure vapors on thin air-cooled surfaces or films.

The operation of all evaporation equipment can be mathematically described by several important basic relationships between total quantity of heat, temperature driving force, mass, physical properties, and type and condition of heat transfer surface.

The following formulas are the most frequently used in the evaluation description of evaporation type equipment. These formulas are described in generally simplified but correct form for easy use and application to these processes.

Heat Transfer Across a Discrete Surface

$$Q = UA \, \Delta T$$

Total heat transferred (Q, Btu per hour) equals heat transfer coefficient (U, Btu per hour per square foot per degree Fahrenheit) times area (A, square feet) times temperature difference (ΔT, degrees Fahrenheit).

Heat Transferred in Heating or Cooling of Fluid

$$Q = w \times C_p \times dT$$

Total heat transferred (Q, Btu per hour), equals mass (w, pounds) times heat capacity (C_p, Btu per pound per degree Fahrenheit) times temperature difference (dT, degrees Fahrenheit).

Heat of Evaporation or Condensation

$$Q = W \times h_{FG}$$

Total heat transferred (Q, Btu per hour) equals mass evaporated or condensed (w, pounds) times heat of evaporation (H_{FG}, Btu per pound).

Heat Transfer Coefficient Across a Discrete Heated or Cooled Surface

$$1/U_{oa} = 1/U_{condensation} + 1/K_{tube} + 1/U_{heating} + f$$

Reciprocal overall heat transfer coefficient (U_{oa}, Btu per hour per square foot per degree Fahrenheit) is equal to the sum of the reciprocal coefficients or resistances adjusted to consistent units of measure.

$U_{condensation}$ represents the film coefficient of steam condensing on one side of the heat transfer surface. K_{tube} represents the resistance due to the metal wall of the tube or transfer surface. $U_{heating}$ is a film coefficient related to the liquid side of the heat transfer surface.

The fouling factor, f, is the resistance to heat transfer caused by the fouling or scale deposition on the heat transfer surface. All units expressed above should be adjusted into unit of Btu's per hour per square foot degree Fahrenheit.

Mathematical relationships involved in typical distillation type desalination plants are centered around the basic equations outlined above. A total mass and energy balance can be drawn around any given stage or effect of an evaporator system.

In the case of multiple stage or multiple effect operations a single stage analogy can still be applied with the feed streams to or the product streams from a stage being the result of adjacent stage operation.

Total energy input into an evaporator is typically the sum of several individual sources. The primary energy source is the motive or driving steam in the first effect or brine heater. In the case of a vapor compression cycle, this is the energy put into the system through the compressor drive.

Secondary sources of energy input from such operations as steam ejector vacuum systems, auxiliary heaters for radiation and venting losses, electrical energy through pumping, and others must also be considered. In single or multiple effect evaporator systems, the primary energy input, Q, is the heat resulting from the condensation of the motive steam.

Mathematically, this would be described as the total mass flow of steam in pounds per hour times the number of Btu's per pound liberated through the condensation and sub-cooling of the liquid phase condensate. Concurrently, this heat is transferred into the saline solution and results in the two steps of heating the feed water to the boiling point and actually boiling to vapor a certain number of pounds of water from the saline solution.

When operating in a multiple effect mode the steam liberated from this effect will be directed to act as the motive steam supply into the next effect in series. In a case of a single-effect operation, steam will be directed to the final condenser where the original motive steam heat energy will be dissipated as warmed condenser water. On an average basis for every 1,000 BTU energy input into the first effect of a multiple effect evaporator system approximately 800 to 900 BTU are transferred on the next effect in series as driving force or motive steam.

Expressed in other terminology, the typical economy for a multiple effect evaporator system is about 0.8 or 0.9 pounds of product water produced per effect for every one pound of motive or input steam energy. Multiple effect evaporator economy is not a consistent relationship in terms of production with each effect added. Because of such factors as boiling point elevation of feed-water, internal losses and pressure drops, deviations from full boiling and others, multiple effect economy is a slowly diminishing function.

The mathematical relationships drawn around a multiple stage flash type system are complicated primarily because of the typically large number of effects. A total mass or energy balance drawn for a single stage of a flash evaporation system is fairly straightforward and easily derived. For a once through system, feed seawater within the condenser tubing section is heated in proportion to the heat given up by the flash brine from the stage.

Additional heat is added via the total number of pounds of steam admitted to the salt water heater. The total quantity of heat available to flash into product water is a function of the total mass of brine flowing and the number of degrees of temperature between the outlet of the salt water heater and the equilibrium pressure within the flash chamber.

This total heat is used to generate vapor based on the relationship that pounds of vapor produced is equal to the total quantity of heat divided by the heat of evaporation per pound. The cycle is completed when the vapor liberated from a stage enters the upper portion of the chamber and gives up heat through condensation to the incoming feed seawater.

Once again, multiple stage or multiple effect operation is a rather complex interplay in which the streams to or from any given stage are the result of the stages on either side. Nevertheless, the basic principles that apply to any one stage will also apply to a multiplicity of stages.

In any evaporation system, the input heat energy into the system must be rejected or discharged at other points in the system. In the case of multiple effect evaporators, the boiler steam heat which was introduced into the first effect is discharged through the heated condenser water and through the product streams.

In simple flash evaporation cycles the steam added to the salt water heater is discharged in both the product distilled water and the waste brine overboard streams. In more complex recirculation type or multiple effect multiple stage flash plants the input energy through the salt water heater is discharged primarily through heated last-stage condenser water or heat rejection water in addition to the standard brine and product distilled water streams. In all cases, irretrievable heat losses occur through radiation and venting from the equipment.

The vapor compression evaporator cycle is more complex from the thermodynamic standpoint. It would properly be described as a adiabatic-isentropic compression cycle. In common words, this means that energy is retained within the system at a constant entropy level. A mechanical energy input to the compressor drive is transferred as heat which results in evaporation of a certain quantity of vapor from the feed water.

The principle of operation of the vapor compression cycle is that saturated vapor at a lower temperature and pressure (frequently atmospheric steam) is directed into the suction side of a compressor. As this steam is compressed heat energy is added and the slightly higher pressure steam is discharged into a tube bundle within the evaporator.

In the tube bundle, the compressor discharge steam condenses, recycling its heat into the feed water, resulting in the vaporization of some water which then becomes the suction feed to the compressor.

Typical compression cycles operate across differentials of five psi or less and with total energy inputs varying between 20 and 100 Btu's per pound of final distilled water produced. One of the unique aspects of the vapor compression cycle is that only one section of heat transfer surface is required. In this heat transfer area motive steam is condensed giving up its heat to the feed water while at the same time this condensed motive steam becomes the final distilled product.

As in many evaporation cycles, auxiliary heat exchangers are utilized to conserve heat between feed, blowdown and distilled product streams. The mathematical description of these heat exchangers is typified by the equations relating to the heating or cooling of fluid as well as to the equation for heat transfer coefficient and heat exchange across a discrete surface.

Sketches of various types of desalination evaporation systems are shown in Figures 1 through 7. A single effect steam motivated still is shown in Figure 1. Input energy is supplied through boiler steam directed into a heating bundle in contact with saline water to be evaporated.

Mass flow of steam in equals mass flow of condensate out, less venting losses. Total heat input into the system is a product of mass flow and the heat given up by condensation of each pound of steam condensed. A minor amount of heat may be contributed due to the sub-cooling of the condensate out of the heating bundle to below the saturation temperature within the bundle. The total Q given up through the heating steam is transferred into the saline feed water. This heat goes into two operations, heating of the feed water and evaporation of a portion of the feed water. Vapors produced from the evaporator bundle are directed into the condenser where their heat is given up in the heating of a specific volume of condenser cooling water. Two product liquid streams in the form of a slightly concentrated brine and pure water distillate are also discharged from the unit. The product of brine and distilled product flow must equal feed water flow. Total heat input through the motive steam must equal the heat added to the condenser water, brine blowdown, and distillate streams.

FIGURE 1: SINGLE EFFECT

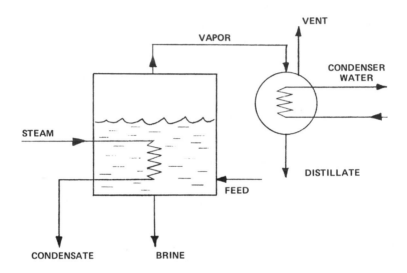

A multiple effect evaporator system is shown in Figure 2. The operation and mathematical description of this unit is identical to a single effect device except that the steam generated from any one effect is directed into the next effect in series.

Likewise, the brine discharge in any one effect becomes the feed to a subsequent effect. Motive steam is condensed in the first effect bundle and is normally returned as clean condensate to the boiler. Condensed steam in all remaining effects and final condenser is the product distilled water from the plant. Once again, total mass and energy balance must be maintained with feed water equaling the total of blowdown and distilled water out. Likewise, heat added through the heating bundle and any other minor sources must be accounted in temperature rise in the condenser water or brine and distillate discharged streams. Remaining minor discrepancies in the mass and energy balance can usually be attributed to radiation and venting losses from the equipment.

FIGURE 2: MULTIPLE EFFECT

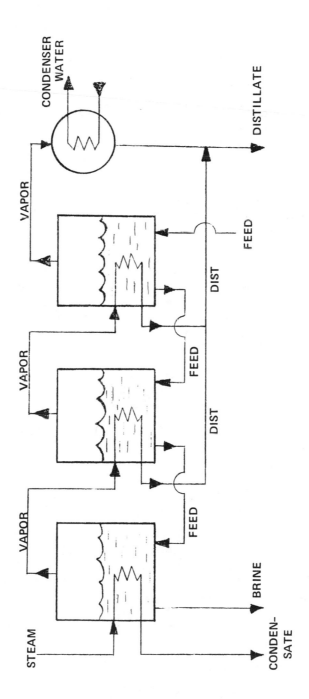

Operation of either single or multiple effect evaporators can be accomplished in many physical forms. Devices such as submerged tube (heating bundle under a pool of saline water), spray film (saline water sprayed over heated tubes), vertical tube (large diameter tubes with rising or falling films of saline water, circulation type (saline water circulated by forced or natural means through either plate or tube type heat exchangers), and several other physical types of evaporators are common desalination units.

The chemical theory and conditions of operation are identical for most types of systems. The practical level for concentration in most units would be slightly below the saturation level of the lowest solubility scale forming material. Scale control through the use of trace amounts of precipitation retarding chemicals such as polyphosphates and polyacrylytes is quite common.

Continuous feed or periodic injection of acid is very effective for the control of alkaline scales. Occasionally complete acid cleaning with soluble salt forming acids is required. On certain types of feed water the use of defoamers and/or marine growth control chemicals are required.

Flash evaporation differs from typical multiple effect evaporation in that the boiling of seawater or other feed water does not occur on a specific heat transfer surface. Figure 3 is a sketch of a simple single effect flash plant. Feed water enters the plant through the condensing tube section commonly in the upper portion of the evaporator. Here the seawater is heated by condensation of the vapors that have been flashed from the brine. Partially heated seawater then enters a salt water heater in which additional heat is added through exchange with a separate source of boiler or other motive steam. Seawater is now at some temperature in excess of the normal boiling point based on the pressure or vacuum within the evaporator shell.

FIGURE 3: SINGLE STAGE FLASH

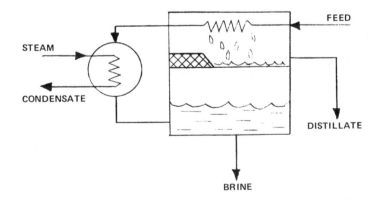

This superheated water reenters the lower portion of the shell through an expansion valve and immediately flashes or gives up a certain amount of heat in the formation of pure water vapor. This vapor enters the upper portion of the shell and is condensed on the outside surface of the cold tubing through which is flowing feed seawater.

In this sense this type of heat exchange is regenerative in that the heat given up in producing the condensed distilled product water is recycled back into the heating of the raw feed water. Mathematically, balance is maintained in that the feed water equals the sum of brine blowdown and distillate product water.

Heat input through the motive steam is discharged in the form of clean condensate and brine and distillate at a temperature higher than that of the feed water. The quantity of distilled water produced by any flash evaporation stage is proportional to the quantity of superheated water entering the stage and the number of degrees of superheat of that water.

In typical single effect flash evaporation plants, the feed water is divided so that approximately 90% is discharged as waste brine while about 10% is recovered as high purity distillate. Because of this relatively low percentage of product water recovered, single effect once through flash evaporation is usually limited to rather economical methods of scale control. As indicated previously, mass and energy balance and the typical equations relating to other evaporation methods also apply to the flash principle.

Figure 4 is a sketch of multiple stage once through flash evaporation cycle. Once again, the basic description of operation is identical to a single stage except that the brine rather than being discharged overboard from any one stage goes on to become the superheated feed to the next stage.

Likewise, the feed water circuit runs in series through all stages up to the salt water heater. Because of the multiple staging principle and the regenerative type of operation higher economies can be achieved than with single stage units. Increase in overall evaporator economy is a function of the number of stages but is a diminishing function as each successive stage is added.

In other words, with all other conditions equal a four stage flash plant is not double the economy of a two stage plant. Likewise, an eight stage plant would be less than double the economy of a four stage plant.

In general, for once through flash evaporation systems, maximum economies achieved are in the order of eight to ten pounds of distilled product water to one pound of input steam energy.

Achieving this goal requires plants of 25 or more individual stages. Percent water recovery with a multiple stage through system is similar to a single stage with approximately 10% of the feed water recovered as distilled product.

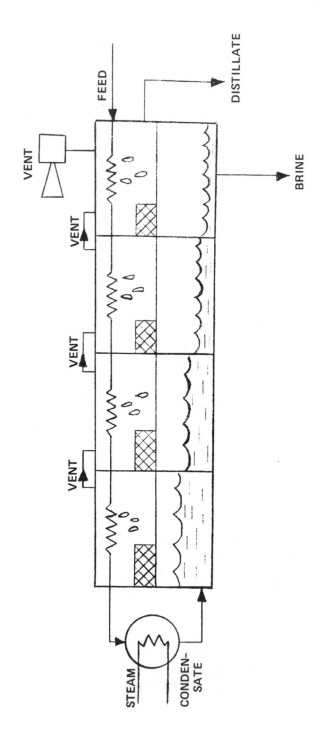

FIGURE 4: ONCE THROUGH MULTIPLE STAGE FLASH

A modification of the multiple stage flash cycle which uses less feed water is the recirculation mode. This operation is indicated in Figure 5. The basic concept of this approach is that a plant is now divided into three major sections. Heat rejection stages are where the feed water entering the tubes is rejected or discharged forming the basic heat sink or heat discharge from the evaporator.

The brine resulting from the last stage of evaporation is recycled into the tube side of an intermediate stage rather than being wasted or directly pumped back to sea. The middle stages of the evaporator are known as the regenerative stages and their function is similar to that of a standard once through plant.

Final heat is added in a salt water heater with superheated brine coming back into the lower portion of the evaporator stages for flashing into high purity product. Depending upon terminal temperature and chemical conditions, it is often possible to recover as much as 50% of the feed water as high purity product in a recirculation type plant.

On this basis, feed water chemical treatments of a more extensive nature can be employed because the relationship of feed water to product water is more favorable.

Recirculation plants often operate with full acidification of feed to prevent alkaline scale formation and at terminal or maximum temperatures for brine in the range of 250°F. Multiple stage recirculation type plants have been constructed with as many as 40 or 50 individual stages.

A most recent innovation of the flash cycle is the multiple effect multiple stage arrangement shown in Figure 6. The multiple effect multiple stage of MEMS system resembles a series of multiple stage recirculation plants overlapping one another.

From a comparison of typical recirculation plant shown in Figure 5 there are three major sections to the plant: heat rejection section, heat regeneration section and salt water heater. If two or more standard plants were placed in a series fashion it would be possible to superimpose the salt water heater section of the lower plant on the heat rejection section of the upper plant. This is precisely what the MEMS arrangement accomplishes.

In the most complex system now in operation three such recirculation plants have been superimposed one upon the other. The result of this procedure is that the center system has no distinct brine heater or heat rejection section. These parts become in fact the heat rejection section of the highest temperature plant as well as the brine heater of the lowest temperature plant.

The largest and most efficient MEMS plant in existence today is constructed of a total of 68 stages in three effects. The capacity of the plant is slightly over one million U.S. gallons per day at an operating economy of 20 pounds of distillate per pound of input motive steam.

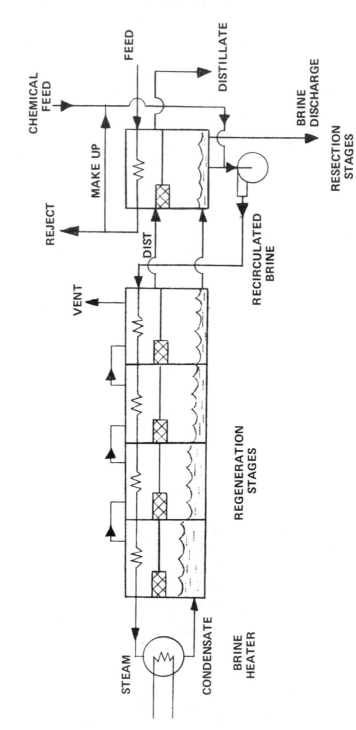

FIGURE 5: RECIRCULATION MULTIPLE STAGE FLASH

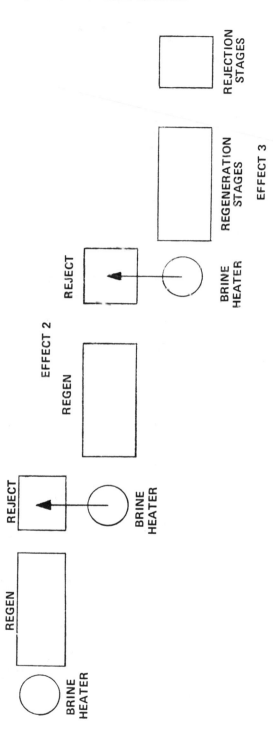

FIGURE 6: MULTIPLE EFFECT MULTIPLE STAGE FLASH

An added advantage exists with the MEMS system in that recirculating brine loops can be maintained as various concentration levels commensurate with optimum scale control conditions.

The typical vapor compression cycle is indicated in Figure 7. This evaporation cycle as indicated previously is adiabatic isentropic compression in which the vapors generated from the evaporator shell are directed into the section side of a compression. In this compressor, mechanical energy is transformed into heat energy and the compressor discharge is directed into the heat input section of the evaporator body.

FIGURE 7: VAPOR COMPRESSION

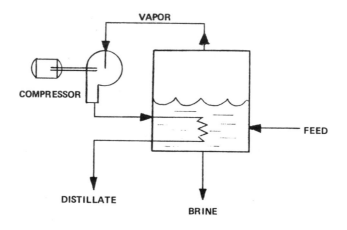

Within the heat input section this motive steam condenses giving up its heat of evaporation to boil the incoming seawater and the condensate becomes the distilled product water. Most vapor compression cycles operate with the shell vented to the atmosphere thereby controlling the suction side of the compressor at approximately 0 psig and about 212°F.

Under these conditions the total heat content of a pound of vapor is approximately 1,100 Btu's per pound. Going through the compressor this steam is compressed to 3 to 5 psig with a saturation temperature of about 220° to 228°F. Under these conditions, the compressed steam has a total heat energy ranging from approximately 1,140 to 1,180 Btu's per pound.

When directed back into the heating side of the evaporator body the total heat of evaporation is given up in condensation. This heat is almost the exact amount needed to evaporate a pound of fresh vapor from the saline water side of the unit. In this cycle the only heat required for maintenance of operation is that heat which represents the difference in temperature level between discharged brine and distillate streams and the incoming feed water and radiation and venting losses from the unit.

General water treatment chemistry, scale control, degasification, and other desalination functions are quite similar for the vapor compression cycle to the other evaporative types of operation. Since most common operation occurs about 212°F. relatively good mechanics for scale control must be utilized. Noncondensible gases pose a severe problem in retarding heat flow, therefore deaeration and degasification of feed waters is a relatively important function in compression cycle operation.

Some practical features which play an important part in the design and operation of vapor compression systems are such things as mechanical maintenance problems on high speed rotating machinery, relative low temperature differences across heat transfer surface, and relatively fixed conditions of operation. Because of these considerations vapor compression evaporators, have been applied to only relatively small production units in desalination.

Standard factory assembled systems up to approximately 25,000 gallons per day are widely used and some partially prefabricated systems up to 50,000 gallons per day have been constructed. In general, plants of larger capacities have had problems with the aspects mentioned above and have not been widely used in the desalination field.

In summary, the mechanical relationships as well as the physical arrangement of evaporation equipment for desalination follows several basic rules and patterns. The quanitative description of heat transfer, mass transfer, and fluid flow applies in common to all evaporation cycles. Physical construction and mechanical means of accomplishing vapor removal from saline water also follow basic patterns independent of the variations between specific evaporation cycles and methods.

DESALINATION BY FREEZING

Paul A. Weiss
Colt Industries Operating Corp.
Water and Waste Management Operation

ABSTRACT

Desalination by freezing is a method of recovering fresh water from sea and brackish waters by means of a liquid to solid to liquid phase change. In that a phase change is involved, it is similar to a distillation system where the separation is accomplished by a liquid to gas to liquid phase change.

The freezing process offers several important potential advantages over a distillation process. Generally less energy is required, corrosion and unsoluble scale problems are reduced, and less expensive materials of construction can be used. Economics and practicality of the freezing process have been demonstrated, but it has not yet been advanced to broad commercial application.

SEPARATION BY FREEZING

The separation of fresh water from a saline solution is based on the rather well known fact of physical chemistry that when a saline solution's temperature is lowered to its freezing point and further heat is removed, the ice crystals which form are pure water. Solutes (dissolved solids) are excluded from the crystal and remain in solution concentrating the remaining brine. Removal of the pure water ice crystal from the concentrated brine, cleansing the crystal of the layer of adhering brine, and melting of the ice produces fresh water.

Desalination by freezing offers a number of advantages over distillation processes. To freeze and thus separate one pound of water requires the removal of only approximately 144 Btu's; the evaporation of one pound of water requires the addition of approximately 1,000 Btu's (depending upon pressure at which evaporation takes place). The energy required to carry on the process is thus potentially reduced. Heat transfer surface requirements are reduced, reducing the cost for these surfaces.

The insoluble scale-forming minerals in sea and brackish waters generally have inverse solubility curves. Their solubility decreases with an increase in temperature and increases with a decrease in temperature. Scale problems in freezing processes are essentially nonexistent since they operate at low temperature. The use of chemicals to control the formation or for removal of scale is therefore not required.

Rates of corrosion from seawater are strongly affected by temperature. With increases in temperature the corrosion rates rise rapidly. The low operating temperatures of the freezing processes greatly reduce the problems of corrosion. Less expensive materials of construction can be used for process equipment reducing their cost.

FREEZING METHODS

Freezing of salt water can be carried out by any number of methods. Any of the methods will result in forming of pure water ice. Selection of the method used is primarily a matter of economics, energy requirements, equipment costs, reliability, ease of operation, and size of the plant.

Freezing on a Metallic Surface

The freezing of the ice can be accomplished directly on a surface cooled by mechanical refrigeration or other method. Ice formed on the surface is removed mechanically by scraping or by alternately freezing and melting on the surface. To produce fresh water the ice is separated from the brine, cleansed and melted. Heat for melting the ice is available as rejected heat from the refrigeration system.

While direct freezing on a surface is straightforward and would appear simple, it has a number of disadvantages. Energy requirements tend to be relatively high; the equipment is complex, expensive, and difficult to operate and maintain. Large metallic heat transfer surfaces are required for both the freezing and melting steps. Further, while the ice formed is pure water, concentrated brine tends to become trapped as pockets between the crystals in the lattice matrix of the ice sheet. This trapped brine is very difficult to remove and results in contamination of the product water when the ice is melted.

Because of the problems with this method, no attempts have been made to use it for commercial production of fresh water. The method has, however, been used for a number of years in the stabilization of wine, a process wherein water is removed from the wine to increase the alcohol and sugar content to improve quality and stability of the wine.

Use of a Secondary Heat Transfer Fluid

By use of a secondary heat transfer fluid for freezing of the ice, the problems of removing the ice from a metallic surface and of brine trapped in the ice can be avoided. In this method the freezing step is accomplished by mixing directly in the salt water a chilled fluid of low miscibility. Kerosene is one of the fluids which have been used. The chilled fluid removes heat from the salt water, freezing a part of the water. The fluid is separated from the salt water, recooled by refrigeration, and recycled to the freezer in a continuous process. The fresh water ice is separated from the concentrated brine, cleansed and melted. Any heat transfer fluid carried with the ice must be stripped from the product water.

While the use of a heat transfer fluid, in place of a metallic heat transfer surface, avoids some of the problems, new problems are introduced. Separation of the fluid from the concentrated brine and product water requires elaborate decanting and stripping equipment. Loss of the secondary heat transfer fluids with the waste brine discharged, even in very low concentrations of a few parts per million, results in an operational cost for replacement so high as to render the process unattractive for the production of water. Failure to completely strip the heat transfer fluid from the fresh water also results in contamination of the product, which can make it unsuitable for human consumption.

No attempt has ever been made, to the writer's knowledge, to produce potable water commercially by means of the secondary heat transfer fluid freeze desalting method. Tests have been conducted using the method to concentrate certain industrial pollution streams. In these tests kerosene was used as the heat transfer fluid. Results of these tests indicated that the probable rates of loss of kerosene in a commercial plant would probably be so high as to render the process uneconomic.

Secondary-Refrigerant Freezing

In the secondary-refrigerant process, freezing of a portion of the salt water feed is accomplished by evaporation of a refrigerant, such as butane, mixed directly with the salt water. The latent heat of vaporization of the refrigerant is provided by the latent heat of crystallization of the water as it freezes. Refrigerant vapor formed in the crystallizer (freezer) is continuously removed to maintain a dynamic steady state condition.

Compression of the refrigerant vapor and contacting the compressed vapor with separated and cleansed ice from the freezer results in condensing of the refrigerant and melting of the ice, the latent heat of the refrigerant providing the heat for melting of the ice. The liquid refrigerant is recycled to the freezer and the melted ice becomes the product water.

Figure 1 schematically illustrates, in simplified, form, a secondary-refrigerant desalting system. The system functions as follows: seawater enters the system through a heat exchanger where sensible heat is removed by heat exchange with cold concentrated brine and cold fresh water product leaving the process. Seawater leaves the heat exchanger near its freezing temperature and enters the freezer where it mixes with the refrigerant. In the freezer a pressure condition is maintained such that the saturation temperature of the refrigerant at that pressure is below the freezing temperature of the saline solution in the freezer, causing a portion of the saline solution to freeze as pure water ice. The latent heat of freezing of the water provides the heat required to evaporate the refrigerant.

FIGURE 1: SCHEMATIC OF A SECONDARY-REFRIGERANT DESALTING SYSTEM

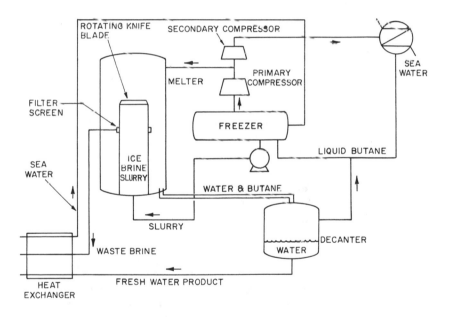

Refrigerant vapors formed in the freezer are compressed by a mechanical compressor maintaining a constant pressure in the freezer. The refrigerant vapor is compressed to a pressure at which the refrigerant's saturation temperature is slightly above the melting temperature of pure water ice.

Brine ice slurry formed in the freezer is pumped to a washer melter column where the ice is separated from the concentrated brine and washed. The cleansed fresh water ice is harvested and transferred to the melter section of the column were it is contacted with the compressed refrigerant vapor from the freezer. The compressed vapor on coming in contact with the ice, condenses, giving up its latent heat and melting the ice.

Fresh water and liquid refrigerant passes from the washer-melter column to a decanter-stripper where refrigerant and water are separated. Liquid refrigerant is recycled to the freezer to form more ice and fresh water is discharged from the system as product. Concentrated waste brine from the washing column is discharged from the system through the heat exchanger.

Heat leaking into the system, inefficiency of the heat exchanger, pump work, work of compression for the refrigerant, etc. all result in more refrigerant vapor being formed than there is ice on which to condense it. This excess vapor from the discharge of the primary compressor is further compressed by a secondary compressor and condensed in a condenser, the heat being rejected to seawater. The liquid refrigerant is recycled to the freezer along with the primary liquid refrigerant stream.

The secondary-refrigerant freezing system enjoys a number of advantages:

(1)   To produce a pound of fresh water requires the transport of only approximately 144 Btu's over a rather small energy level gradient, thus energy requirements of the system are small.

(2)   Bulk of the heat transfer takes place across phase barriers rather than through metallic heat transfer surfaces, and heat transfer surface requirements are therefore small.

(3)   By proper selection of the refrigerant the system can be operated at or near ambient pressures.

(4)   Problems from insoluble scale and corrosion are minimal.

Balancing the advantages of the system are some formidable technical and economic problems. Refrigerants such as butane, which are most attractive from a point of cost, availability and thermodynamic characteristics, tend to form a waxlike substance called hydrates under process conditions. These hydrates seriously interfere with the ice washing process and the complete stripping and recovery of the refrigerant from the waste brine and product water streams. The lost refrigerant results in high operating costs for replacement and possible contamination of the product water. Storage of refrigerant also poses a problem since storage for the entire system charge must be provided if process is shut down for any reasons.

Extensive research and pilot plant studies have been carried out in development of the secondary-refrigerant process. This effort continues with at least one company working with fluorocarbons (Freon 114, 115 and 318) as a refrigerant in an attempt to advance the process to commercialization. As yet no secondary-refrigerant commercial plants for production of potable water have been installed in the free world.

Vacuum-Freezing Vapor-Compression Process

The vacuum-freezing vapor-compression desalting process is similar in some respects to the secondary-refrigerant process. In the VFVC process part of the process water acts as the refrigerant. Use of water eliminates the problems of costs, supply, contamination, separation, etc. of the third process fluid, refrigerant.

The primary principles involved for converting of saline water to fresh water in the VFVC process are rather well known in physical chemistry and are summarized below.

(1)   When saline water is boiled, the vapor that is produced is pure water; the salts are concentrated in the remaining brine.

(2)   When saline water is frozen, the individual ice crystals consist of pure water; the salts are concentrated in the remaining brine. However, each crystal is coated with a layer of concentrated brine that adheres to the surface of the crystal, and this layer must be removed by washing.

(3)   The freezing point of typical saline water is essentially unaffected by reductions in pressure.

(4)   The boiling point of saline water varies drastically with the pressure. By reducing the pressure to about 3.9 to 4.6 mm. of mercury absolute, the boiling point of the saline water is reduced until it is the same as the freezing point. In other words, at a pressure of 3.9 to 4.6 mm. of mercury, depending on the dissolved solids concentration, the saline water can boil and contain ice simultaneously.

(5)   To convert one pound of water into vapor about 1,070 Btu must be applied. To convert one pound of water into ice, 144 Btu must be removed. The ratio of the heat that is added in producing one pound of vapor to the heat that is removed in producing one pound of ice is about 7.5 to 1.

(6)   The melting point of pure ice is $32.0°F.$, and the vapor pressure of the ice is the same as that of pure water at the same temperature. The pressure is 4.58 mm. of mercury.

The principles discussed above, applied to a practical VFVC desalting system, are illustrated in the schematic flow diagram, Figure 2. This simplified diagram shows both the essential fluids in the system.

FIGURE 2: SCHEMATIC FLOW DIAGRAM OF VFVC DESALTING SYSTEM

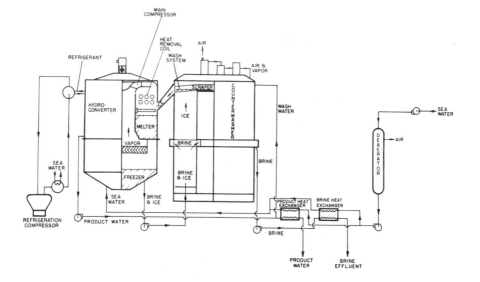

Saline water is pumped at ambient temperature from the source through a filter to remove entrained solids. This water then flows through a vacuum deaerator that removes the dissolved gases that could otherwise interfere later with the heat transfer in the melter.

From the deaerators, the now-deaerated saline water flows into the system through heat exchangers where the saline water is cooled by heat exchange with the cold brine and the cold product water flowing separately out of the system. This deaerated saline water, now cooled almost to its freezing point, is then pumped into the freezer.

Under the influence of the low pressure in the freezer (about 3.4 mm. of mercury), the saline water boils so that part flashes into vapor. In so doing the vapor extracts its heat of vaporization from the rest of the saline water since no external heat is supplied to cause the boiling. But the bulk of the saline water is already at its freezing point. Therefore, the removal of additional heat from this cold saline water causes a portion of it to freeze, and to give up its heat of crystallization. About 7.5 pounds of ice are formed for each pound of vapor.

Depending on the salinity of the saline water between 30 and 95% of the saline water feed is converted into pure ice and vapor. Because the salts in this converted portion of the saline water are left behind, the remaining saline water becomes more and more concentrated until it contains 5 to 7% of salts.

The slurry, or mixture, of brine and ice crystals is removed continuously from the freezer and is pumped into the counterwasher. Here the ice crystals are propelled countercurrently to the stream of wash water distributed on the top of the ice pack, and the ice crystals are washed free of all adhering brine solution. The washed ice, now free of salts, is scraped via a chute into the melter.

The water vapor produced in the freezer by the simultaneous boiling-freezing process is removed from the freezer by the vapor compressor, is compressed to an absolute pressure of 4.8 mm. of mercury, and is discharged into the melter. At a pressure of 4.8 mm. of mercury the vapor condenses when brought into contact with the washed ice. As the vapor condenses it gives up its heat of vaporization, and this heat is absorbed by the ice in melting. The condensed vapor and melted ice form cold product water.

The heat transfer in the freezer, during the formation of the ice and vapor, takes place at the surface of the saline water; and in the melter the heat transfer, as the vapor condenses and ice melts, takes place at the surface of the ice.

The cold product water from the melter and the cold brine effluent from the counterwasher are discharged through the heat exchangers to cool the incoming saline water. However, some heat still enters the system with the feed saline water. Also, the energy of the pumps and compressors utilized in the process shows up as heat, and some heat enters the system through the insulation. To maintain thermodynamic balance a refrigerated coil in the melter is used to condense the excess vapor resulting from the loss of ice due to this heat input.

Because the operating pressure is well below atmospheric, some air inevitably enters the system. This air is removed continuously by a combination blower-condenser-vacuum pump.

The VFVC process has a number of advantages, the most important ones being:

(1)   Low energy requirements, typically in the range of 45 to 50 kwh per 1,000 U.S. gallons of fresh water produced from seawater.

(2)   No process chemical requirements, eliminating the cost and operational problems associated with such chemicals.

(3)   Low feedwater and cooling water requirements.

(4)   No scale or corrosion problems.

(5)   Minimum metallic heat transfer surfaces.

(6)   Common inexpensive materials of construction such as low carbon steel and aluminum.

(7)   No potential pollution problem — waste brine discharges from the system at a temperature only a few degrees F. from that of the feed water and does not contain any process chemicals or copper ions.

Colt Industries has carried on extensive research and field testing of the process. Field tests have been conducted at three locations: a 250,000 U.S. gpd plant at Eilat, Israel; a 100,000 U.S. gpd unit on St. Croix, U.S. Virgin Islands; and a 60,000 U.S. gpd pilot plant at the Office of Saline Water, Research and Development Test Station, Wrightsville Beach, North Carolina.

Equipment is not presently being offered on a general commercial basis but development and equipment improvement work continues and it is expected will be made available commercially in the future.

Vacuum-Freezing Ejector Absorption Process

Colt Industries Water and Waste Management Operation presently has under development a new freeze desalting process called vacuum-freezing ejector absorption. The work is being performed under an R&D contract with the Office of Saline Water, U.S. Department of the Interior.

In this process the function of compression of the water vapor formed in the freezer is accomplished by means of an absorber and low pressure steam ejector rather than a mechanical compressor. It offers promise of further reducing energy requirements and the capability of using inexpensive low gradient steam rather than relatively expensive mechanical energy. The practical size of VFVC plants is limited by the design limits of the mechanical compressor. The VFEA system does not suffer such limitations and appears suitable for plants of almost any size, even hundreds of millions of gallons per day.

Figure 3 depicts schematically one possible arrangement of a vacuum-freezing ejector absorption plant. The system would function as follows:

Feedwater enters the system through a cold water deaerator which removes the dissolved gases, reducing the dissolved oxygen concentration to less than 1 ppm. Deaerated feedwater is pumped through three fluid heat exchangers where the feedwater is cooled by heat exchange with cold concentrated brine and cold product water leaving the process. The feedwater leaves the heat exchanger at about 33°F. and enters the freezer.

Pressure in the freezer is maintained at slightly more than 3 mm. Hg absolute. This is below the vapor pressure of the solution at its freezing temperature and part of the feedwater rapidly vaporizes. Since the heat of vaporization must come from the water, part of the water freezes giving up its latent heat of crystallization. An ice brine slurry of approximately 17% mass ice fraction is formed. To minimize concentration gradients and facilitate vapor release, the fluid in the freezer is mechanically agitated.

FIGURE 3: VACUUM-FREEZING EJECTOR ABSORPTION PLANT

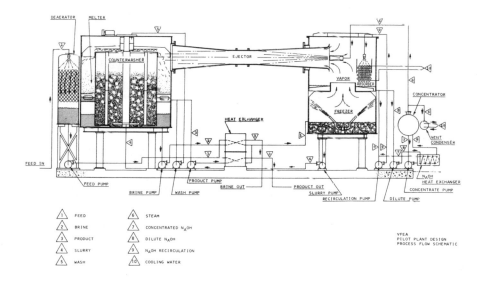

Brine ice slurry is pumped from the freezer to the counterwasher. In the counterwasher the ice consolidates into a porous mass. The brine entering the counterwasher in greater volume than the ice must travel through the ice to the brine drainage screens. This relative motion of the brine to the ice creates a hydraulic pressure gradient which supports the column of porous ice and propels it continuously upward. Part of the cold fresh water produced by the process is distributed over the top of the rising drained ice column. This fresh water passing down the column relative to the rising ice, displaces the brine surrounding the ice crystals, cleansing the ice.

The brine-freshwater interface is controlled by regulation of the brine drainage back pressure. By proper design of the counterwasher, loss of fresh water for washing can be held to less than 5% of the gross water produced.

Brine drained from the ice is pumped from the counterwasher and discharged from the process through the heat exchanger. Fresh water is harvested from the top of the counterwasher and transported to the melter section.

Approximately 38% of the water vapor released in the freezer is absorbed on a 49% solution of sodium hydroxide in the absorber. The heat of absorption is removed by cooling seawater circulated through the cooling coil over which the sodium hydroxide is sprayed. The dilute sodium hydroxide solution, now at a concentration of about 44%, is pumped from the absorber through a heat exchanger where it picks up heat from hot regenerated absorbent from the concentrator.

In the concentrator the dilute absorbent is heated to 250°F. by direct contact with steam. This drives the absorbed water off at a pressure of 300 mm. Hg, regenerating the absorbent. Regenerated sodium hydroxide is recycled to the absorber.

The 250°F. 300 mm. Hg steam from the concentrator is used as the motive fluid for the vapor ejector. The ejector compresses the vapor formed in the freezer which is not absorbed in the absorber. The vapor compressed to about 5 mm. Hg absolute is discharged to the melter along with the motive steam. At this pressure the saturation temperature is above the melting temperature of the fresh water ice. The vapor condenses on the ice melting the ice. The combined condensed water vapor and melted ice becomes the cold product water which is discharged from the system through the heat exchanger.

SUMMARY

For a number of years freezing processes have been recognized as being one of the most promising methods for economic production of fresh water from sea and saline sources. Great effort on the part of many people has been devoted to the development of these processes and important advances have been achieved. Full commercialization has not yet been realized, but it appears the time may be near where freeze desalting plants for economic production of fresh water will be available.

A CASE HISTORY OF A WATER DISTILLATION PLANT
ISLAND OF ANTIGUA

Carl W. Miller
Former Manager
Hawksbill Beach Hotel
Island of Antigua

I would like to discuss with you my experience as a water distillation plant operator while managing the Hawksbill Beach Hotel at Antigua. The Hawksbill is a 50 room resort hotel located on the leeward side of the island and depends on three main sources for its water supply; rain, government, and a 600 gpd MECO distillate plant.

To begin my discussion, I would like to cover three two-part points: (1) Why and how our hotel came to operate a distillation plant; (2) our normal procedures and problems with the plant; and (3) improvements that I think should be made with distillation plants and the future of them with a small business operator.

WHY AND HOW WE GOT THE PLANT

In the Spring and Summer of 1968, as most of you may know, Antigua experienced a severe drought. During this period in order to keep the hotel open, we had to buy and truck water to our reservoir at a cost of $20 to $25 (U.S.) per 1,000 gallons. With a usage rate of 200 gal./room/day or 10,000 gal./day total, it isn't hard to imagine why we were looking for something to cut down our water bill.

It was at this time that the Mechanical Equipment Company of New Orleans convinced eight hotel operators of Antigua that we should buy their 600 gpd, diesel driven, seawater distillation plant. By the Fall of 1968 the hotels were so desperate that instead of waiting for surface shipment we had the plants flown in from Miami. This was no small cost when you consider their 800 pounds plus weight.

By Christmas of 1968 our plant was in operation following a cash commitment of over $50,000. This included the costs of shipment, plant buildings, pumps, settling tanks, and a line to the sea. A considerable outlay for a small business, but well worth it, if it allowed us to survive. And then the fun began.

NORMAL PROCEDURES AND PROBLEMS

To assist us and the other plant operators, MECO placed a field representative on the island to help train our mechanics and ease into the operation as smoothly as possible. They sent a fine man and an excellent mechanic. He spent time with our chief mechanic checking him out in its operation and doing his best to train him in its safety and maintenance.

I'm afraid that most of his fine points fell on deaf ears, as our mechanic's biggest projects to date had been lawn mowers and jeep engines. He learned to start the plant, run it, clean it, and shut it down well enough, but to tune it, balance it, and carefully perform preventative maintenance, was asking too much of him.

Our mechanic also deceived us by letting on that he knew more about its operation than he actually did and he reported more preventative maintenance than was performed. For example, the plant requires a sulfuric acid cleaning every 24 to 48 hours of operation. Our men hated to do this due to its inherent danger. (I could hardly blame them).

I learned later that our mechanic allowed the plant to operate continuously for over 6 days without cleaning. This of course required that the evaporator dome be removed and the perulator tubes bored out. So more expenses; both to us and MECO. Another time the evaporator dome vacuum relief valve malfunctioned. The mechanic didn't notice the low pressure until the dome collapsed. More lost time and expense.

In only 5,000 hours of operation, the diesel engine had to be overhauled three times. It is a good Allis-Chalmers engine too; I blame it on poor preventative maintenance. More lost time and expense.

I did a cost study on the plant for its first two years of operation and we found that it cost us $7.75/1,000 gallons of distillate produced. This compared to the present government water cost of $3.00/1,000 gallons, the supply of which over these two years had improved greatly with the end of the drought. Now MECO claims, and I agree with them, that the cost of distillate should be $2.50 to $2.75/1,000 gallons. But the replacement of parts cost kills this claim when you have to depend on the quality of labor force available in Antigua.

Plant noise also disturbs the serenity of a resort. MECO makes an electric powered plant, which I'm sure would have been much quieter, but at the time, the government electricity was so poor and undependable that a self-powered unit was our only choice. You make do with what you have. But there were many advantages of distillate to us too. We could advertise absolutely pure drinking water, which is of great concern to a tourist. And it really does taste good compared to rain water and government piped water, and the beautiful crystal clear ice cubes for the bar had a benefit too.

The ice machine and washing machine nozzles didn't clog up with mineral deposits. But for all its advantages, the cost of its operation has caused us to put it in a semi-mothball status. We run it enough to maintain it, but no longer for a water source. It's there if we should need it, and we hope that it will help us survive should another drought occur.

IMPROVEMENTS NECESSARY FOR A FUTURE WITH SMALL OPERATORS

The major drawback to our plant in our environment is its excessive number of moving parts. As you well know, if a part moves, it will likely break, so we need help in cutting down the cost of replacement parts.

For example, the 600 has 5 high speed V-belts and 8 low speed belts, three centrifugal water pumps, two three-way float valves, a high speed compressor, 12 hand valves, a contained reciprocating engine, a clutch assembly and innumerable elbows and joints, all of which will eventually break or leak, usually sooner than later if they handle hot salt water which is occasionally sweetened with 2 gallons of concentrated sulfuric acid.

So all you have to do is design something that will produce 1,000 gal./min., have no more than five or six moving parts, is easy to maintain, can be operated by a trained chimpanzee, and yet sell for under $25,000.

If you can solve some of the problems mentioned, you will have a great future with small operators. Otherwise we are not able to afford the present expensive, complicated machinery except in situations where our survival is at stake.

DESALINIZATION
BY REVERSE OSMOSIS 1970
by J. McDermott

The need for fresh water in many arid and highly industrialized areas of the world is rapidly becoming acute. The Office of Saline Water of the U.S. Department of the Interior is actively involved in developing and evaluating various techniques for the desalinization of sea and brackish waters. One of the processes showing most promise to date is reverse osmosis.

For water-purification, reverse osmosis is very attractive since it potentially involves a very low energy consumption compared to other techniques such as distillation and freeze-concentration processes which involve a change of state.

This book summarizes the patent literature through June 1970 relating to reverse osmosis processes with particular emphasis on membrane technology and equipment design. The accompanying Table of Contents provides a clear picture of material covered by this review.

209 pages

DESALINIZATION
BY FREEZE CONCENTRATION 1971

by J. McDermott

This volume is the second of a series of three, dealing with the major desalinization processes. A previous book described reverse osmosis, and a future book will describe distillation techniques. Although less attention has been paid to freeze desalting processes in the United States, active development is being undertaken in England and Japan with promising results.

The need for better potable water supplies is well known. Desalinization represents a major hope for production of adequate water supplies. In addition, a study carried out by the Office of Saline Water (OSW) for the Department of the Interior, pointed out the potential for industrial by-product recovery using desalinization methodology.

The OSW believes the potential market for desalinization equipment for public water supply will grow to probably well over $1 billion before 1980. To this can be added a sizeable market for desalinization equipment applied to industrial waste water. Clearly, the information contained in this volume will provide needed know-how concerning the renewed interest in freeze concentration processes.

The Table of Contents below indicates the total coverage of freeze concentration processes and equipment offered by this volume. The numbers in ( ) indicate where more than one process has been covered in a particular area.

DESALINIZATION BY DISTILLATION 1971
RECENT DEVELOPMENTS
by J. McDermott

This volume concerning distillation processes for desalting water, is the last in a series of three books dealing with the major desalinization processes. These three books will provide you with comprehensive coverage of technological advances in this field, which is destined to be a major source of concern for the next few decades.

The prospect of revitalizing the arid and semi-arid areas of the world, together with the problem of providing increased quantities of potable water for the already developed areas of the world has already produced considerable research in desalinization processes.

In the past two years alone, over one hundred patents relating to distillation processes have been issued in the U.S. Multistage, multieffect flash distillation processes are continually being improved to provide better utilization of energy, and to control scale formation. Research in vapor compression distillation has already led to an operational plant.

This book reviews the recent patent literature of distillation processes. The Table of Contents below indicates the coverage this text provides.

CORROSION RESISTANT MATERIALS HANDBOOK 1971

Second Edition
by I. Mellan

This famous book, first published in 1966, has doubled in size in its new edition. This increase in size (without an increase in price) is necessitated in part by our environmental problems. Corrosion, always an urgent and persistent problem, bothers and baffles us even more today, because of the quantity and complexity of chemicals in our polluted biosphere. The fast pace of technology has greatly increased the amount of waste material cast into the atmosphere from industrial stacks and domestic chimneys, and into rivers, lakes and oceans which receive dumped industrial wastes, agricultural runoff from the soil, and organic human wastes.

Metals have always been subject to corrosion, but are currently under even greater attack from chemical pollutants such as pesticides, hydrocarbons, alkalis, sulfur dioxide, automobile exhaust chemicals and tobacco smoke, to name but a few. Although metals have been the most obvious target of corrosive attack by atmospheric chemicals, other materials also are susceptible to chemical change. The great complexity of modern life demands the use of new materials resistant to this erosion which is active everywhere. Each new advance in the physical and chemical sciences carries with it new problems of corrosion.

The chemical industry is particularly susceptible to corrosive influences such as fumes, vapors, dusts, contaminants, by-products and water. Some industries have special corrosive problems characteristic of them. Examples of such industries are the manufacture of acids, such as hydrofluoric, sulfuric, phosphoric, and acetic, the manufacture of chlorine and caustic soda, synthetic rubber, insecticides, synthetic fibers and metal processing.

In the light of present knowledge, the term "corrosion prevention" is interpreted as the means by which corrosion is minimized or lessened and not by the necessity for preventing it entirely. There is no container or coating that will be absolutely immune to all chemicals under all conditions. Any material chosen for a specific purpose requires a knowledge of its physical, chemical, thermal and mechanical properties; its resistance to corrosion; its workability; its serviceability; and its cost. Although cost is important, it should not be the prime consideration.

The ideal metal, alloy or plastic coating, would be one that is entirely impervious and resistant to all corroding media. Since no such material exists an economic compromise must be made. In the past, corrosion resistance was associated with a few metals. This has changed, however, and today a choice may be made from among new alloys, wood, glass, ceramics, new plastics and resins.

This book will help you cut losses due to corrosion by enabling you to choose the proper commercially available corrosion resistant materials for your particular purpose. Most projects involve such a high capital expenditure that only tried and accepted design methods are applicable.

The value of this book lies in the extensive double indexing of thousands of chemicals. The 147 tables are arranged by type of corrosion resistant material. The index is organized by corrosive chemicals and refers you to specific recommendations in the tables. The compounds covered also represent the newer products and trends in corrosion resistant materials.

The tables in this book were selected by the author from manufacturers' literature at no cost to, nor influence from, the manufacturer of the materials.

Contents:

490 pages.

ENVIRONMENTAL SCIENCE TECHNOLOGY INFORMATION RESOURCES 1973

Edited by Dr. Sidney B. Tuwiner in conjunction with
The Chemical International Information Center
(Chemists' Club, New York)

Environmental science is a new science in many of its aspects. Being a subject of high actuality, it has led to a proliferation of many new publications, stemming from a pressing need for continual updating of the technology and its literature, codes regulating discharges and emissions, etc.

Environmental science is interdisciplinary, involving sociology, law and economics, as well as technology. Some of the problems arising from this fact are discussed here and information retrieval solutions are discussed.

The individual papers give a close look at the many sources of environmental information from the several viewpoints of librarian, editor, and specialist in government sources or industry associations, enabling the user to obtain a broad perspective on environmental information.

Section I includes the proceedings and panel discussions of the Symposium on Environmental Science Technology Information Resources sponsored by the Chemical International Information Center, and held at the Chemists' Club, New York, N.Y. on April 28, 1972.

Introduction — Dr. Sidney B. Tuwiner

Resources and Industrial Cooperation on Environmental Control at the American Petroleum Institute — E. H. Brenner, American Petroleum Institute

The Environmental Protection Agency and its R & D Program — William J. Lacy, Chief, Applied Science & Technology, Office of Research & Monitoring, EPA

Government Sources of Information on Environmental Control — Marshall Sittig, Director of Special Projects, Princeton University

Sorting It Out - Data vs. Information — Steven S. Ross, Editor, New Engineer

Environmental Control Resources of a Large Technical Library — Kirk Cabeen, Director of the Engineering Societies Library

Section II includes selected papers presented at the National Environmental Information Symposium, sponsored by the EPA, held at Cincinnati, Ohio, September 24-27, 1972.

Technical Information Programs in the Environmental Protection Agency — A. C. Trakowski, Deputy Asst. Administrator for Program Operations, Office of Research & Monitoring, EPA

The Environmental Science Information Center — James E. Caskey, Director ESIC (Environmental Science Information Center)

Federal Environmental Data Centers and Systems — Arnold R. Hull, Associate Director for Climatology & Environmental Data Service, NOAA, U.S. Dept. of Commerce

Scientific and Technical Information Centers Concerned with the Biological Sciences — William B. Cottrell, Director, Nuclear Safety Information Center, Oak Ridge National Library

Secondary Technical and Scientific Journals — Bernard D. Rosenthal, President, Pollution Abstracts, Inc.

Environmental Litigation as a Source of Environmental Information — Victor John Yannacone, Jr., Attorney

Scientific and Technical Primary Publications Carrying Environmental Information — D. H. Michael Bowen, Managing Editor, Environmental Science & Technology, American Chemical Society

Applications of Socioeconomic Information to Environmental Research and Planning — William B. DeVille, Director of Program Development, Gulf South Research Institute

Socioeconomic Aspects of Environmental Problems: Secondary Information Sources — James G. Kollegger, President, Environment Information Center, Inc.

Section III is a bibliography of basic governmental, institutional, and organizational documents assembled by the United Nations Conference on the Human Environment, held at Stockholm, Sweden, June 5-16, 1972.

Introduction
Basic Documents Received from States Invited to the Conference
Basic Documents Prepared Within the United Nations System
Basic Documents Received from Other Sources

ISBN 0-8155-0467-5

218 pages

ANTIFOULING MARINE COATINGS 1973

by A. Williams

Coatings Technology Review No. 1

One of the earliest needs for performance-oriented coatings was in the marine environment. Very early formulas were designed around known toxins such as copper and mercury compounds, and the earlier patent literature is replete with hundreds of directions for using these materials in creosote and natural drying oil formulations. Later information indicated in this book involves more sophisticated materials.

The two areas of a ship requiring specialty coatings are the bottom and the boot-topping area. The boot-topping area, intermittently exposed to air, sunshine, and water, represents a surface particularly difficult to protect from the elements.

For ships' bottoms, antifouling compounds, based on copper, mercury, and tin, are incorporated into somewhat water-sensitive binders to afford gradual breakdown of the film to allow for a sustained release of the poison. This required self-erosion necessitates frequent repainting of the ship's bottom, depending on geographical location and severity of exposure. By contrast, boot-topping paints are designed to provide a high level of resistance to both salt water and weather. Typically phenolic resin-tung oil, vinyl resin combinations are used.

This book describes many patented processes which provide high performance antifouling coatings based on metal compounds as well as organic coating compositions.

A partial and abbreviated table of contents is given here. Numbers in ( ) indicate the nos. of patents or application techniques discussed under a given heading. Chapter headings are given, followed by examples of important subtitles.

ISBN 0-8155-0464-0

271 pages

DESCALING AGENTS
AND METHODS 1972
by J. A. Szilard

Scale, in the context of this review, is either the deposit from hard water (calcium sulfate or carbonate, magnesium carbonate, iron oxide or carbonate, silica or silicates, or other particles originally present in the water system) or it is heat generated scale on metals (usually their oxides). Scale deposits from organic liquids, such as milk, are also encountered.

Historically, water treatment went through three phases. First there was the lime-soda process to precipitate and remove the hardness-causing materials. Next came the ion exchange resins. Both of these methods were stoichiometric in nature. Then, in the forties, the "threshold effect" was discovered. This involves the addition of substoichiometric amounts of descaling agents, at times as low as 1 to 2 parts per million, to prevent the deposition of scale, to alter the crystal form of the precipitate, if it forms; or to generate a thin, protective film on the substrate to prevent deposits from adhering. Such threshold activity is shown by polyphosphates, polyamines, and polyelectrolytes in general.

190 patent-based processes. The numbers in ( ) indicate the numbers of processes or application techniques under a given topic. Chapter headings are given, followed by examples of important subtitles.

1. TREATMENT OF WATER SYSTEMS (51)
Phosphates, Organic Phosphor Compounds, and Chromates (25)
Polyamines, Polyamides, and Other Polyelectrolytes (9)
Acids and Their Derivatives (8)
Tannins and Lignins (4)
Sulfur Compounds (3)
Silicates (1)
Silicones (1)

2. BOILER WATER TREATMENT (35)
Boiler Scale Treatment with Organic Acids and Derivatives (14)
  Alginic Acid or Alginates
  Alginic Acid and Sodium Nitrite
  Mixtures of Sulfamic Acid and Sodium Nitrite
  Mixtures of Sulfamic and Carboxylic Acids with Ammonium Inhibitor
  Organic Acid Solutions with an Inhibitor
  Formic and Citric Acid Mixtures

Sodium Citrate and Sulfonated Lignin
Alkali-Modified Cannery Waste Product
Nitrilotricarboxylic Acid Salts
Gluconic Acid
Polycarboxylic Acid Complexes with Corrosion Inhibition
Polyacrylonitrile
Carboxylic Acid and Sodium Nitrite
Sodium Nitrilotriacetate and Styrene Maleic Anhydride Polymer
Ethylene-Maleic Acid Copolymers
Thiourea and Other Sulfur Compounds (9)
Morpholine and Amine Combinations (3)
Phosphates and Phosphites (3)
Phenols (2)
Plant Materials (2)
Chromates (1)
Peroxides (1)

3. OIL WELL EQUIPMENT PROTECTION (51)
Phosphates and Other Phosphorus Compounds (16)
Polyamines and Derivatives (12)
Sulfur Compounds (5)
Lignins and Gluconates (4)
Sodium and Ammonium Salts (4)
Cellulose and Starch Derivatives (3)
Alkalis, Alcohols, Phenols (7)

4. SEAWATER AND BRINE CONVERSION (17)

5. MILL SCALE (11)
Alkaline Solutions (4)
Inhibited Acids (4)
Boron Compounds (2)
Sodium Hydride (1)

6. FOOD INDUSTRIES (9)
Dairies and Breweries (6)
Sugar (3)

7. NUCLEAR REACTORS (4)

8. VARIOUS SCALE PREVENTION MEANS (12)
Ammonia-Soda Process (1)
Textile and Paper Production (4)
Soap and Cosmetics Manufacture (3)
Aqueous Cooling Systems (2)
Steam Irons (1)
Rust Removal (1)

237 pages